Full Frontal Calculus

Full Frontal Calculus

Seth Braver

South Puget Sound Community College

Cover photo by Shannon Michael.
(In Latin, *calculus* = “small stone”.)

ISBN-13: 9781081888909

This book is occasionally updated with minor changes. Most recent update: 6/22/21.

Contents

Acknowledgements

When this book was a work in progress, I hid it on an invitation-only website that kept track of those who visited it. I was pleased to see a former colleague of mine, Hank Harmon, now at Georgia Gwinnett College, appearing regularly on its log of visitors – even several years after he had left SPSCC. Part of my reason for making this book publicly available was the thought that if someone of Hank's intellectual caliber found it valuable, then surely a few other heterodox mathematics professors would concur. A deep bow then to Hank for his unwitting encouragement from afar.

Cesar Villasana, another ex-colleague, was unfailingly enthusiastic about my textbook projects. He even wrangled faculty support for my failed bids for a sabbatical in which I had proposed to complete this book's prequel, *Precalculus Made Difficult*. Cesar's last name is apt: Before he "retired" from SPSCC (the active voice is not quite right for that verb), his quiet integrity and fundamental kindness made his office a uniquely "healthy village" in a land ravaged by bureaucratic mediocrity.

Yet another ex-colleague, Melissa Nivala (now – God help her – at Evergreen State College), had nothing at all to do with this book, but I thank her anyway. Melissa has already been acknowledged in the well-known differential equations textbook by Blanchard, Devaney, and Hall, but that was some time ago; it's high time she be acknowledged again. I am not one to forget the precious gift of a freshly-picked lobster mushroom from the forest in one's own backyard.

Thanks to my current colleagues for enjoying (or at least tolerating) my peculiarities and heresies. Thanks to my students who read and commented on early drafts of the book, catching all manners of typos, mistakes, and infelicitous expressions therein. Thanks to Arthur Schopenhauer, Philip Larkin, Hermann Melville, and Franz Schubert for many hours of stimulation and consolation. Thanks to Olympia's great blue herons, wood ducks, mallards, mourning doves, and bald eagles.

Thanks to The Pooka.

Above all, thanks to Shannon, who couldn't care less about calculus, but without whom my life would be so radically different that this book probably wouldn't exist.

Preface for Teachers

"Anyone who adds to the plethora of introductory calculus textbooks owes
an explanation, if not an apology, to the mathematical community."
-Morris Kline, *Calculus: An Intuitive and Physical Approach*

Full Frontal Calculus covers the standard topics in single-variable calculus, but in a somewhat unusual way. Most notably, in developing calculus, I favor infinitesimals over limits. Other novel features of the book are its brevity, low cost, and – I hope – its style.

Why infinitesimals rather than limits? A fair question, but then, why limits and not infinitesimals? The subject's proper historical name is *the calculus* ***of infinitesimals***. Its basic notation refers directly to infinitesimals. Its creators made their discoveries by pursuing intuitions about infinitesimals. Most scientists and mathematicians who comfortably use calculus think about it in terms of infinitesimals. Yet in our classrooms, we hide the essence of calculus behind a limiting fig leaf, as if fearful that the sight of bare infinitesimals could make fragile young ladies faint at their desks. Please. Spare us your smelling salts. The tightly corseted pre-Robinsonian era ended half a century ago.

Infinitesimals can now be made every bit as rigorous as limits, though full rigor is beside the point in freshman calculus, the goals of which are deep intuition and computational facility, each of which we enhance by admitting infinitesimals into our textbooks. By using infinitesimals, we help our students relive the insights of the giants who forged the calculus, rather than those of the janitors who tidied up the giants' workshop.

The book is short because it need not be long. If fools like you and me (and Silvanus Thompson) can master the calculus, it cannot be so complex as to require a 1000-page instruction manual.

I have chosen self-publication primarily because my experiences with a conventional publisher for my first book, *Lobachevski Illuminated*, were not altogether happy. Despite winning the Mathematical Association of America's Beckenbach Prize in 2015, that book passed most of its first decade in awkward electronic limbo – until 2021, when the American Mathematical Society gave it its first proper print run. By retaining control over the present book's publication, I can ensure that it remains available to my students (and yours) both in a free electronic version and an inexpensive print-on-demand paper version.

For the past several years, while using drafts of *Full Frontal Calculus* in my classes at South Puget Sound Community College, I've offered students extra credit points for catching typos. Their quarry, once plentiful, has now been satisfactorily reduced. Any remaining typos or errors are, of course, due solely to my students' appalling negligence. Please let me know (through my campus email) of any surviving mistakes you might notice. And if, after reading the book, you find yourself clamoring for full frontal *multivariable* calculus, let me know that, too. Given sufficient demand, who knows, I might even write it.

Preface for Students

"Allez en avant, et la foi vous viendra."
-Jean Le Rond D'Alembert

Fair warning: This book's approach to calculus is slightly unorthodox, particularly in its first two chapters. Consequently, you are unlikely to find other books or videos that will show you how solve many of its homework problems.[*] This, despite what you might initially think, is a good thing; it will force you, from the beginning, to become an active reader – and thus to become more intellectually self-reliant.

Full Frontal Calculus is meant to be read slowly and carefully. Ideally, you should read the relevant sections in the text before your teacher lectures on them. The lectures will then reinforce what you've understood, and clarify what you haven't. Read with pencil and paper at the ready.[†] When I omit algebraic details, you should supply them. When I use a phrase such as "as you should verify", I am not being facetious. Only *after* reading a section should you attempt to solve the problems with which it concludes. When you encounter something in the text that you do not understand (even something as small as an individual algebraic step), you should mark the relevant passage and try to clear it up, which may involve discussing it with your classmates or teacher or reviewing prerequisite material. Regarding this last point, calculus students commonly find that their grasp of precalculus mathematics is weaker than they had supposed. Whenever this happens ("I've forgotten how to complete the square!"), do not despair, but *do* review the relevant topics.[‡] Be assiduous about repairing foundational cracks whenever you discover them, for the machinery of calculus is heavy; attempts to erect it on porous precalculus foundations end poorly.

As with learning a language or a musical instrument, learning calculus requires tenacity of purpose: To succeed, you must devote several hours a day to it, day after day, week after week, for many months. Fortunately, the intrinsic rewards in learning calculus – as with a language or the violin – are substantial. And, of course, knowing calculus extends your ability to study science, and thus eventually to enter professional scientific fields.

At the beginning, this can all seem quite intimidating. Bear in mind that tens of thousands of people succeed in learning basic calculus every year. You can be one of them. But it will require hard work, and at times, you may wonder whether it is worthwhile. It is. Go on, and faith will come to you.

Let us begin.

---

[*] Once we've reached cruising altitude (Chapter 3), I, your faithful pilot, will switch off the fasten-seatbelts light, and you'll be able to consult standard calculus texts and websites once again if and when you so desire.

[†] Although you can read this book for free online, I recommend that you purchase a print copy so that you can scrawl notes in the margins. Online reading, at least in my experience, is rarely active reading.

[‡] See this book's prequel, *Precalculus Made Difficult*, available on my website, BraverNewMath.com.

Part I
Differential Calculus

Chapter 1
The Basic Ideas

Calculus: From A to The

A calculus is a set of symbolic rules for manipulating objects of some specified type. If you've studied statistics, you've probably used the *calculus of probabilities*.* If you've studied formal logic, it follows that you've met the *propositional calculus*.† After you've mentally summed up all the little bits of knowledge in this book, you'll have learned **the calculus *of infinitesimals***.

An infinitesimal is an infinitely small number – smaller than any positive real number, yet greater than zero. Like square roots of negatives (the regrettably-named "imaginary numbers"), infinitesimals seem paradoxical: They manifestly do *not* belong to the familiar system of real numbers. Mathematicians and scientists have, nonetheless, used them for centuries, because infinitesimals help us understand *functions of real numbers* – those faithful tools with which we model, among other things, the so-called real world of experience. To gain perspective on our real homeworld, it helps to survey it from without. Such a justification for working with numbers beyond the reals should satisfy the hard-nosed pragmatist; for another, equally valid, justification, we need only observe our fellow mammals the dolphins and whales at play. There is pleasure to be had in breaching the surface of a world that normally confines us.

In biology, the naked eye is perfectly serviceable for some purposes, but a microscope's lenses reveal details of a microworld that can ultimately help explain what we experience on our familiar scale. Similarly, mathematicians have found that although the real numbers suffice for our basic measurement needs, "infinitesimal-sensitive lenses" can sharpen the pixilated image that the reals present to our naked mind's eye. Our mathematical microscope is, of course, purely mental. Cultivating a sense of what it reveals requires practice and imagination, but the essence of the idea, however, is simple: Magnitudes that appear equal to the naked eye (i.e. in terms of real numbers) may turn out, when viewed through our infinitesimal-sensitive microscope, to differ by an infinitesimal amount. Conversely, magnitudes that differ by a mere infinitesimal when viewed under the microscope correspond, when viewed with the naked eye, to *one and the same real number*.

"Interesting," you may reflect, "but I wish we'd just stick with the good old familiar real numbers." Be careful what you wish for, lest you unnecessarily limit your mathematical imagination! Yes, the reals are familiar and indispensable, but they can also be disturbingly strange; by embracing infinitesimals, we can actually divest the reals of some of their strangeness, as you'll see in the first example below. In the second example, you'll see how infinitesimals can help us bridge the qualitative divide between curves and straight lines. Bridging this divide turns out to be a major theme of calculus.

Example 1. Suppose an urn contains eight balls, one of which is red. If we draw a ball at random (so each ball is equally likely to be drawn), the probability of drawing the red one is clearly $1/8$. But suppose there are not eight, but *infinitely many* balls from which to draw. Under these circumstances, what is the probability of drawing the one red ball?

The answer seems to be "one in infinity", but what does that even mean? Well, there are two things we can definitely say about a "one in infinity" probability. The first is that it is surely *less than* "one in N" for any whole number N whatsoever. The second is that, like all probabilities, it is a

* The calculus of probabilities contains rules such as $P(A \cup B) = P(A) + P(B) - P(A \cap B)$.

† Such a calculus would contain rules such as $P \wedge Q \Rightarrow P$.

number between 0 and 1, the values that correspond to impossibility and certainty respectively. Combining these two facts yields the following conclusion: The probability of drawing the red ball is some number that lies between 0 and 1 and is less than $1/N$ for all whole numbers N.

A little thought should convince you that only one *real* number satisfies both demands: **zero**. Hence, *if we confine ourselves to the real numbers*, we are forced to conclude that the probability of drawing the red ball is zero – which suggests that drawing it is not merely unlikely, but actually impossible. Moreover, the same logic applies to each one of the infinitely many balls, which leads us to the awkward conclusion that from our infinite collection, it is impossible to draw any ball whatsoever! Such is the paradoxical scene as viewed with the naked eye.

The paradox vanishes if we accept infinitesimals. For if we do, we need not conclude that the probability of drawing the red ball is zero; it could be *infinitesimal* while still meeting the two demands for a "1 in ∞" probability described above. The probability of drawing the red ball would then be unfathomably minuscule – less than $1/N$ for every whole number N, impossible to represent as a decimal, indistinguishable from zero in the *real* world – and yet, not quite zero. Consequently, the red ball *can* be drawn, though I wouldn't advise betting on it. ♦

So much for balls. Let's drop down a dimension and discuss circles. Everyone and his mother "knows" that a circle of radius r has area πr^2. But *why* is this so? Infinitesimals can help you understand.

Example 2. The area of a circle with radius r is πr^2.

Proof. Inscribe a regular n-sided polygon in the circle. Clearly, the greater n is, the closer the polygon cleaves to the circle. Even when n is relatively small (say, $n = 50$), distinguishing the two shapes is difficult for the naked eye, yet they remain distinct for any finite n. Our proof of the area formula, however, hinges on a radical reconceptualization: *We shall think of the circle as a regular polygon with infinitely many sides, each of which is infinitesimally small.* This idea will enable us to use facts about polygons (straight, simple objects) to learn about circles (curved objects).

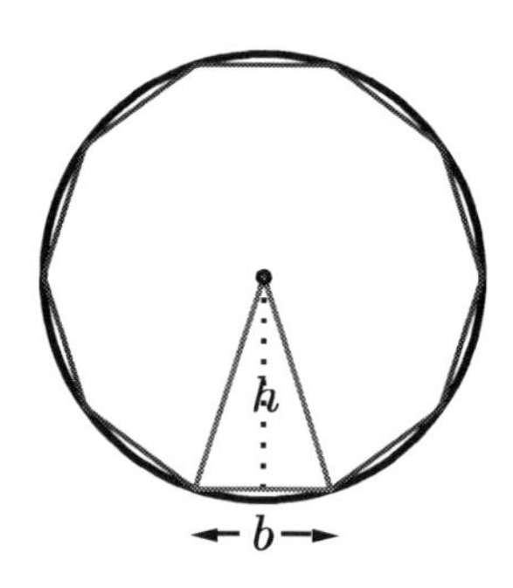

Since a regular n-gon's area is n times that of the triangle in the figure above, its area must be $n(bh/2)$. Since $nb = P$ (the polygon's perimeter), this expression for area simplifies to $Ph/2$.

Thus the circle, *being a polygon*, has area $Ph/2$. But for our circle/∞-gon, P represents the circle's circumference (which is $2\pi r$), and h represents the circle's radius (which is r). Substituting these values into the area formula $Ph/2$, we conclude that the *circle's* area is $(2\pi r)r/2$, which simplifies to πr^2, as claimed. ■

Please dwell on this surprising, beautiful, disconcerting argument. When mathematicians began to use infinitesimals, even philosophers and theologians took note. Is a proof that uses infinitesimals a genuine proof? Is a circle really a polygon of infinitely many sides? Is it wise for finite man to reason about the infinite? Even as such philosophical debates raged (from the 17th century on), mathematicians paid only halfhearted attention, busy as they were developing a potent *calculus* of infinitesimals. That this calculus worked no one questioned; that it lacked a fully comprehensible foundation no one denied. Its triumphs were astonishing. The infinitesimal calculus helped physics break free of its static Greek origins and become a dynamic modern science. And yet... all attempts to establish iron-clad logical foundations for the calculus failed. Since at least the time of Euclid (c. 300 BC), mathematics had been viewed as the archetype of logical reasoning. Small wonder then that, despite the undeniable utility of the infinitesimal calculus, its murky basis received stinging criticism. Most famously, philosopher George Berkeley suggested in 1734 that anyone who could accept the mysterious logical foundations of the infinitesimal calculus "need not, methinks, be squeamish about any point in Divinity." To compare mathematics – the traditional rock of logical certainty – to theology, nay, to assert that mathematicians, far from proceeding by perfectly rigorous thought, "submit to authority, take things on trust, and believe points inconceivable" (as Berkeley would have it) was to shake one of the very pillars of Western civilization.*

Not until the late 19th century did mathematicians discover a perfectly rigorous method (the theory of *limits*, which you'll learn about in Chapter π) to set the theorems of the infinitesimal calculus on solid foundations. That it took so long to develop these foundations is understandable, given the surprising sacrifice involved: To transfer the massive body of theorems onto the long-desired secure logical foundations, mathematicians had to sacrifice the infinitesimals themselves!

Placed firmly atop these new limit-based foundations, the many theorems of calculus that had been developed over the previous centuries were finally secure (there was no longer anything to fear from the philosophers), but the infinitesimals that had nourished the subject as it developed were ruthlessly expunged in the victory celebration. The very notion of an infinitesimal came to be viewed as an embarrassment to the brave new limit-based calculus, as though "infinitesimal" were a discredited religious idea from a more primitive time whose abandonment was necessary for the further progress of humanity. Even the subject's name was changed. What had once proudly been known as "The Calculus of Infinitesimals" thenceforth became known simply as **The Calculus**, a name whose emptiness spoke – to those, at least, with ears to hear – of the ghosts of departed quantities.†

Infinitesimal ghosts continued to haunt the calculus, for although the theory of limits had brilliantly disposed of a logical problem, it had introduced a psychological problem. In the minds of many who used calculus as a scientific tool (but who had no particular concern for the subject's esoteric logical foundations), infinitesimals remained far more intuitive than limits. For this reason, much notation that originally referred to infinitesimals was, surprisingly, retained even after the great infinitesimal purge. Naturally, the notation was reinterpreted in terms of limits, which entailed a sort of mathematical schizophrenia. One would use infinitesimal notation and think infinitesimally, but good mathematical hygiene demanded that one refrain from actually mentioning infinitesimals in public. To be sure, textbook authors would sometimes timidly advise their readers that it might be helpful to think of such and such an expression in terms of infinitesimals, but such advice was invariably followed (as if an authority figure had just returned to the room) by a stern warning that actually, infinitesimals don't exist, and one shouldn't speak about such things in polite society.

---

* Berkeley was a masterful shaker of pillars. His denial that matter exists outside of minds paved the way for David Hume's philosophical demolition of causality and personal identity, and hence to Immanuel Kant's subsequent reconstruction of these ideas on a radically new philosophical basis (transcendental idealism) that he developed to refute Hume, and which then became a cornerstone of modern philosophy.

† Since "The Calculus" is only slightly more expressive than "The Thing", there was no real loss when, in time, even the definite article was shed. Hence today's unadorned *Calculus* (on tap at a college near you).

Resurrection

"I think in coming centuries it will be considered a great oddity in the history of mathematics that the first exact theory of infinitesimals was developed 300 years after the invention of the differential calculus."
- Abraham Robinson

In his landmark book *Nonstandard Analysis* (1966), from which the preceding epigraph was taken, Abraham Robinson astonished the mathematical world by using tools from 20th-century logic to construct the Holy Grail of Calculus: a perfectly rigorous way to make infinitesimals logically respectable.

Robinson used his newly vindicated infinitesimals (which joined the familiar *real* numbers to produce an extended system of *hyperreals*) to construct an alternate foundation for calculus. His infinitesimal-based, 20th-century foundation was every bit as solid as the limit-based, 19th-century foundation, but the mathematical world, in the intervening years, had grown secular; its desire for the Infinitesimal Grail had been nearly extinguished. Though duly impressed by Robinson's intellectual achievement, mathematicians were largely unmoved by it, for the theory of limits (itself a century old by that time) was not only fully rigorous, but also fully entrenched. The logical foundations of calculus had long since ceased to be an active field of inquiry, and few mathematicians (who were busy, naturally, exploring other problems) cared to revisit it. Their position was quite understandable: The theory of limits, which they, their teachers, and their teachers' teachers had all mastered long ago as mere undergraduates, *worked*. From a strictly logical standpoint, it did not need to be replaced, and mathematicians are no less inclined than others to hearken unto the old saw: *If it ain't broke, don't fix it.* Thus they tended – and tend – to view Robinson's work as a remarkable curiosity. All mathematicians know of Robinson's achievement. Few have studied it in detail.

O intended reader of this book, you are not a professional mathematician. You are a student in a freshman-level class. You need not, at this point in your academic career, concern yourself with the full details of calculus's logical foundations, except to be reassured that these exist and are secure; you can study them (in either the limit-based version or the infinitesimal-based version) in the appropriate books or classes should you feel so inclined in the future. One does not take a course in Driver's Education to learn the principles of the internal combustion engine, and one does not take freshman calculus to learn the subject's deep logical underpinnings. One takes a course such as this to learn the calculus itself – to learn what it is, how to use it, and how to think in terms of it, for calculus is as much a way of thinking as it is a collection of computational tricks. For you (and for your teacher), the importance of Robinson's work lies not in its formidable logical details, but rather in the retrospective blessing it bestows upon the centuries-old tradition of *infinitesimal thinking*, a tradition that will help you understand how to think about calculus – how to recognize when calculus is an appropriate tool for a problem, how to formulate such problems in the language of calculus, how to understand *why* calculus's computational tricks work as they do. All of this becomes considerably easier when we allow ourselves the luxury of working with infinitesimals. We need no longer, as in the 1950's, blush to say "the i-word". And so, in accordance with this book's title, infinitesimals shall parade proudly through its pages, naked and unashamed.

The World of Calculus: An Overview

"I'm very good at integral and differential calculus,
I know the scientific names of beings animalculous.
In short in matters vegetable, animal, and mineral,
I am the very model of a modern major general."
- Major General Stanley, in Gilbert and Sullivan's *Pirates of Penzance*.

Calculus is traditionally divided into two branches, integral and differential calculus.

Integral calculus is about mentally decomposing something into infinitely many infinitesimal pieces; after analyzing the pieces, we then re-*integrate* them (sum them back up) to reconstitute the whole. The spirit of integral calculus hovered over our proof of the circle's area formula, when we reimagined the circle's area as the sum of the areas of infinitely many infinitesimally thin triangles. Integral calculus thus involves a special way of thinking, or even a special way of seeing. With "integral calculus eyes", one might view a solid sphere as a stack of infinitely many infinitesimally thin discs, as suggested by the figure at right. Alternately, one might imagine it as an infinite collection of concentric infinitesimally thin hollow spheres, nested like an onion's layers.

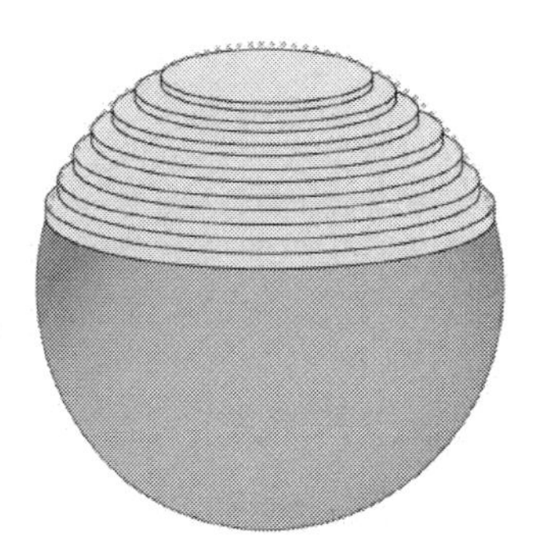

Differential calculus is about *rates of change*. An object's speed, for example, is a rate of change (the rate at which its distance from a fixed point changes in time). A bank's interest rate is a rate of change (the rate at which dollars left in a savings account change in time). Chemical reactions have rates of change (the rate at which iron rusts when left in water, for instance). The slope of a line is a rate of change (the rate at which the line rises as one runs along it). Where there is life, there is change; where there is change, there is calculus. Differential calculus is specifically concerned with rates of change on an *infinitesimal* scale. Thus, it is not concerned with how the temperature is changing over a period of weeks, years, or centuries, but rather with how the temperature is changing at a given *instant*.

Naturally, the two branches of calculus work together: To understand large-scale global change, we mentally disintegrate it into an infinite sequence of local instantaneous changes; we scrutinize these infinitesimal changes with differential calculus, and then we re-*integrate* them with integral calculus, so as to see the whole again with new eyes and new insights.

In this book, we'll begin with differential calculus, and then move on to integral calculus.

Differential Calculus: The Key Geometric Idea

Differential calculus grows from a single idea: **On an infinitesimal scale, curves are straight**.

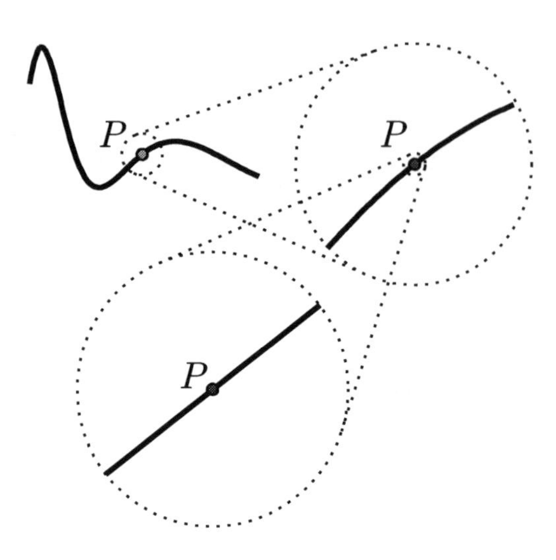

To see this, imagine zooming in on a point P lying on a curve. As you do so, the part of the curve you can see (an ever-shrinking "neighborhood" of P) becomes less and less curvy. In an *infinitesimally* small neighborhood of P, the curve coincides with (part of) a straight line.

We call this straight line the curve's **tangent at *P***. Thus, in the infinitesimal neighborhood of a point, a curve and its tangent are indistinguishable. Outside of this neighborhood, of course, the curve and its tangent usually go their separate ways, as happens in the figure at left. Nonetheless, simply by recognizing that curves possess "local linearity", we can answer seemingly tricky questions in geometry and physics. Consider the following examples, in which tangent lines make unexpected appearances. No calculations are involved, just *thinking* in terms of infinitesimals.

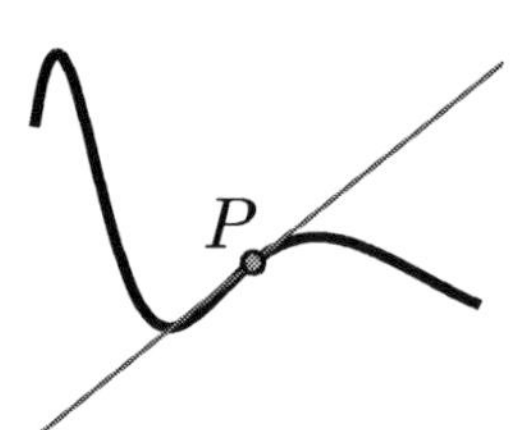

Example 1. (An Illuminating Tangent on Optics)

Light reflects off flat surfaces in a very simple manner: Each light ray "bounces off" the surface at the same angle at which it struck it. This much has been known since at least Euclid's time (c. 300 BC). What happens, however, if the surface is *curved*? We shall reason our way to the answer.

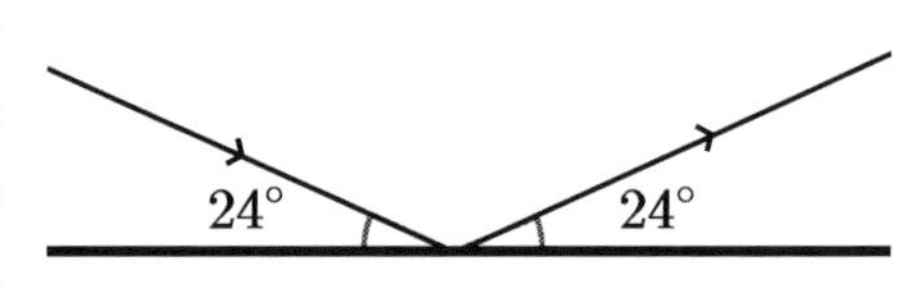

First, note that in the case of a flat surface, we only need an infinitesimal bit of surface against which to measure the relevant angles. (Erasing most of the surface, as in the figure, clearly leaves the angles unchanged.) As far as the light ray is concerned, most of the surface is redundant. The ray's angle of reflection is a strictly local affair, determined entirely in an infinitesimal neighborhood.

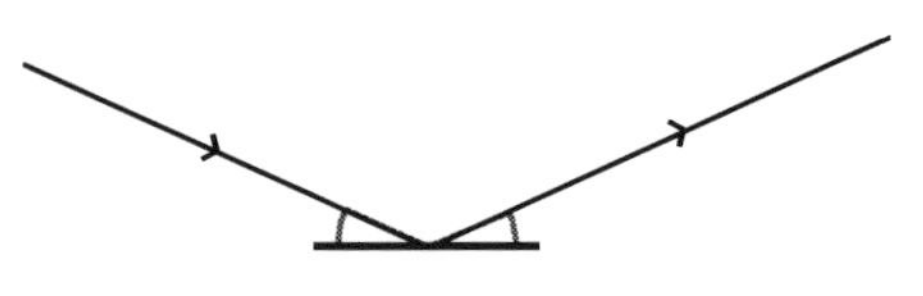

So what happens when a light ray strikes a *curve*? Think locally! Let us mentally visit an infinitesimal neighborhood of the light ray's point of impact. There, the curve coincides with its (straight) tangent line, and our curvy conundrum disappears: At this scale, the curve *is* straight, so the old rule still holds! Newly enlightened, we zoom back out to our usual perspective, extend the tangent line as in the figure at right, and know that *this* is the line against which we should measure our reflection angles. ♦

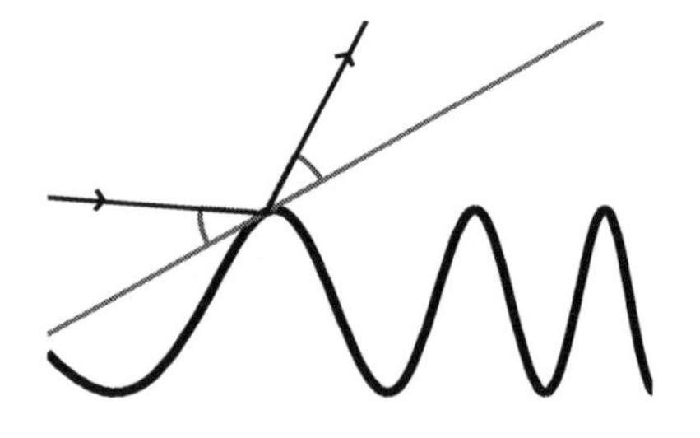

Example 2. (So Long and Thanks for All the Fish.)

If the Sun were to vanish, where would the Earth go? Isaac Newton taught us that the Earth is kept in its orbit by the Sun's gravitational force. He also taught us (following Galileo) that when no forces act on an object moving in a straight line, the object will continue moving along that line. But what about an object moving on a *curved* path that is suddenly freed from all forces that had been acting on it? Well, if we think infinitesimally, we recognize that during any given instant – any infinitesimal interval of time – an object moving on a curved path is actually moving on a *straight* path. Thus when the Sun vanishes, Earth will continue to move along the straight path on which it was traveling in that very instant. That is, the Earth will move along the *tangent* to its elliptical orbit. ♦

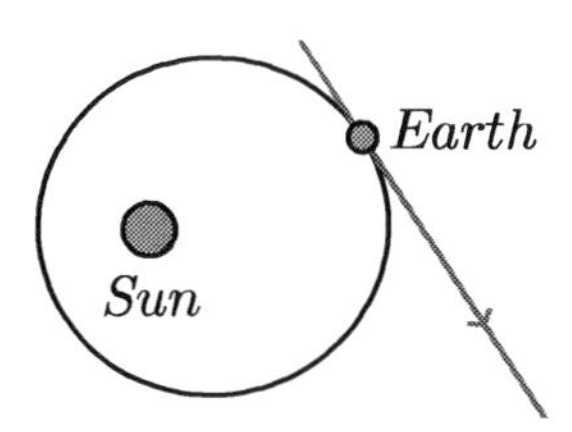

Tangents, in short, are important. Let us pause for a few exercises.

Exercises.

1. Let P be a point on a straight line. Describe the tangent to the line at P.

2. Let P be a point on a circle. Describe the tangent to the circle at P. [Hint: Consider the diameter with P as an endpoint. The circle is symmetric about this diameter, so the tangent line at P must also be symmetric about it. (Any "lopsidedness" of the tangent would indicate an asymmetry in the circle – which, of course, doesn't exist.)]

3. a) When two curves (not straight lines) cross, how can we measure the *angle* at which they cross? Explain why your answer is reasonable.

b) How large is the curved angle between a circle and a tangent to the circle? (cf. Euclid, *Elements* 3.16.)

4. If you stand in the open country in eastern Washington, the earth looks like a flat *plane* (hence "the plains"). Of course, it isn't flat; you are actually standing on a sphere. Explain why the earth looks flat from that perspective, and what this has to do with the key idea of differential calculus.

5. Some graphs lack tangents at certain points. Explain why the graph of $y = |x|$ lacks a tangent at its vertex. [*Hint: Reread the first few paragraphs of this section.*] The gleaming calculus machine does what it was designed for phenomenally well, but it was not built to handle corners. At corner points (which, fortunately, are rare on the graphs of the most commonly encountered functions), differential calculus breaks down.

Rates of Change

Once we recognize the local straightness of curves, it affects even the way we think about *functions*. On an infinitesimal scale, any function's *graph* is a straight line, which, in turn, is the graph of a *linear* function. Hence, **on an infinitesimal scale, all functions are linear!***

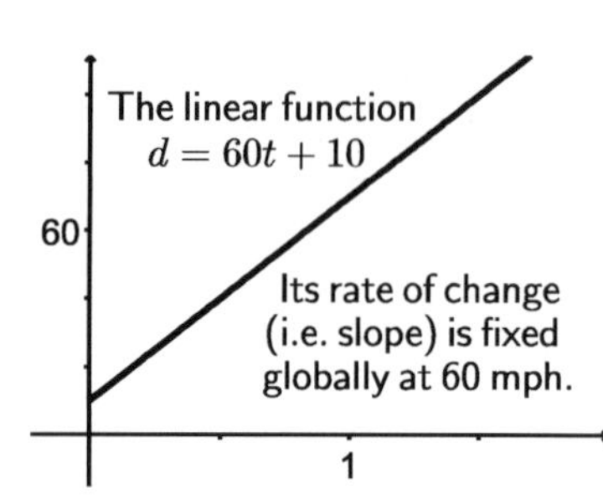

This is excellent news, for linear functions are baby simple. They are simple because the output of any linear function *changes at a fixed rate*, which we call its *slope*. ("Slope" and "rate of change" are more or less synonymous.) If, for example, your car is $60t + 10$ miles from your house t hours after we start a stopwatch, then its distance from your house is changing at the fixed rate of 60 miles per hour. English has a word for the rate at which distance changes in time, so we might as well use it: Your car's *speed* is fixed at 60 miles per hour.

Few phenomena we wish to study are so obliging as to conform to a strict linear relationship. No one actually drives in accordance with the linear function in the previous paragraph – at least not for any appreciable length of time. Rather, a driver's speed varies from moment to moment, even if, on average, he drives 60 miles per hour. But even though physical phenomena are rarely linear in a *global* sense, the functions with which we model them are still *locally* linear. This is precisely why linear functions are so important: Linear functions underlie all functions.

Accordingly, it makes sense to speak of a nonlinear function's *local* rate of change or slope. This notion of a *local* (also called *instantaneous*) rate of change is familiar to every driver; when you glance down at your speedometer, its reading of "63 mph" indicates that your car is moving down the highway at that particular rate *at that particular instant*. Even a few seconds later, your speed may differ. When a policeman aims his speed gun at your car, he is measuring your *instantaneous* rate of change.

To sum up: Locally, the curved graph of a nonlinear function coincides with a straight tangent line. **The tangent's slope is the function's *local* slope, which is the function's rate of change near the point of tangency.** With this insight, we may visually estimate a nonlinear function's local rate of change. For example, consider the figure at right, which is a more realistic *nonlinear* graph of a car's distance travelled (in miles) as a function of time (in hours). The slopes of the two tangents shown in the figure measure the distance function's local rate of change (i.e. the car's speed) at two different times. The graph thus shows us quite plainly that the car was moving faster after one hour than it was after half an hour.

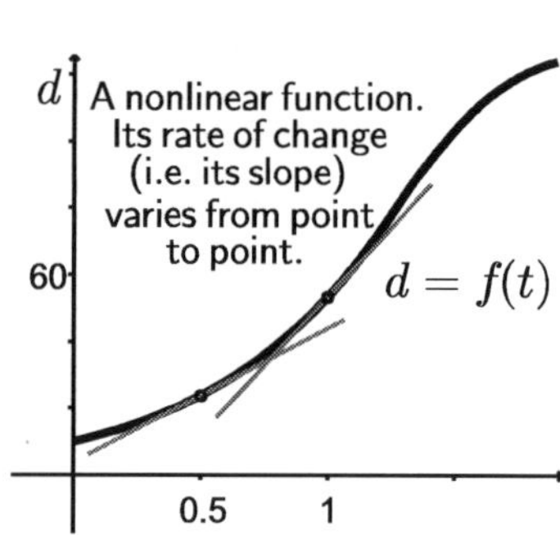

---

* Exceptions to this rule occur in infinitesimal neighborhoods of "corner points" of the sort described in exercise 5. For the sake of readable (if slightly inexact) exposition, I shall continue to make universal statements about "all functions", trusting the intelligent reader, who has been forewarned, always to bear in mind that calculus breaks down where no tangent line exists.

The Derivative of a Function

We are finally ready to define the central object of differential calculus, the **derivative** of a function.

> **Definition.** The **derivative** of a function is a new function whose output is the original function's *local slope* (or equivalently, *local rate of change*) at the given input value.
>
> If f is a function, its derivative is denoted $\boldsymbol{f'}$ (which we read "f-prime").
>
> (There are other notations for a function's derivative, which you'll meet in due time.)

This idea is best understood through a few examples.

Example 1. If a function f is given by the graph at right, then according to the derivative's definition, $f'(0)$ measures the graph's *slope* (i.e. the slope of its tangent) when $x = 0$. Clearly, in an infinitesimal neighborhood of $(0, -1)$, this graph resembles a line whose slope is approximately 1, so

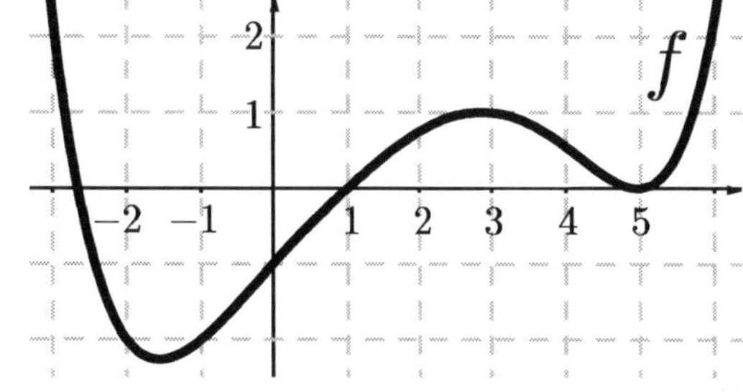

$$f'(0) \approx 1.$$

Similarly, we see that $f'(5) \approx 0$, and $f'(4) < 0$. ♦

Example 2. Suppose that after driving for t hours, Jehu's car has travelled a total of $f(t)$ miles. By the derivative's definition, $f'(t)$ represents the *rate* at which his distance is changing at time t. That is, $f'(t)$ represents Jehu's *speed* (in miles per hour) at time t.

If, for example, $f'(1/2) = 112$, then we know that exactly 30 minutes after he began driving, Jehu was driving furiously (112 mph). If, three hours later, traffic brought his car to a standstill, then we have that $f'(7/2) = 0$.

A graph of Jehu's distance function f would have a slope of 112 when $t = 1/2$, and a horizontal tangent line when $t = 7/2$. ♦

Most people use "speed" and "velocity" synonymously, but these have distinct meanings in physics, where "speed" is just a magnitude (a positive number), while "velocity" is a magnitude *with a direction*. When we analyze motion in one dimension, we use positive and negative numbers to specify direction. For example, when we consider just the vertical motion of a ball (disregarding its horizontal motion), a velocity of – 5 ft/sec signifies that the ball is *descending* at 5 ft/sec.

In such contexts, positive and negative numbers also extend the notion of ***distance*** *from a point* (e.g. "3 meters away") to the richer concept of ***position*** *relative to a point* (e.g. "3 meters *to the right*"). For example, if a ladybug paces back and forth on a line, one point of which we call the origin and one direction of which we deem positive, we might have occasion to describe her *position* as $+8$ inches at one moment, and -8 at another, depending on which side of the origin she happens to be. (In both cases, her *distance* from the origin is 8 inches.)

With these distinctions in mind, we can say that the rate of change of *position* with respect to time is *velocity*. Consequently, the *derivative* of an object's position function describes the object's velocity.

Example 3. When Eris tosses a golden apple in the air, its height after t seconds is $h(t)$ meters. Since h is the apple's *position* function (in the up/down dimension), its derivative, $h'(t)$, gives the apple's *velocity* (in m/s) after t seconds.

For example, if we are told that $h(1) = 12$ and $h'(1) = 5$, we may interpret this as follows: After one second, the apple is 12 meters high and rising at 5 m/s. If we also learn that $h(2) = 12$ and $h'(2) = -5$, we know that *two* seconds after leaving Eris's hand, the apple is once again 12 meters high, but it is now on its way down, for it is *descending* at 5 m/s.

When the apple reaches its zenith, its velocity must be *zero*: At that instant, it is neither ascending nor descending, but *hanging*.

Thus, if the apple's maximum height occurs when $t = 3/2$, we'd have $h'(3/2) = 0$. ♦

Velocity is the rate at which position changes. *Acceleration* is the rate at which *velocity* changes. Hence, we measure acceleration in units of velocity per unit time, such $(m/s)/s$ (abbreviated, alas, as m/s^2).* Note the chain of three functions linked by derivatives: The derivative of an object's *position* function is its *velocity*; the derivative of its velocity function is its *acceleration*.

Acceleration is thus the *second derivative* of position, where "second derivative" simply means "the derivative of the derivative". The second derivative of a function $f(x)$ is denoted, unsurprisingly, $f''(x)$.

Example 4. If we neglect air resistance, any body in free fall (i.e. with no force but gravity acting on it) near Earth's surface accelerates downwards at a constant rate of 32 ft/s^2. Consequently, if $f(t)$ describes the height (in feet) of such an object, and t is measured in seconds, we immediately know that the second derivative of f is a constant function: $f''(t) = -32$. ♦

Position and velocity are rare – but vitally important – examples of functions whose derivatives have special names. Even when such names do not exist, you can always interpret any given derivative as the rate at which its output changes with respect to its input.

Example 5. Suppose $V(t)$ represents the volume (in m^3) of beer in a large vat, where t is the number of hours past noon on a certain day. Throughout the day, beer leaves the vat (as people drink it), but new beer is also poured in by Ninkasi, the Sumerian beer goddess.

Here, $V'(t)$ represents the *rate at which the volume of beer in the vat is changing at time t*. If, say, $V'(5.5) = -0.5$, then at $5:30$ pm, the vat's volume is decreasing at a rate of half a cubic meter per minute. (Even Ninkasi struggles to satiate thirsty Sumerians after the 5 o'clock whistle.) Horrors! Will the beer run out? Well, suppose we also learn that $V(6) = 0.0001$ and $V'(6) = 2$. These values imply that at $6:00$, the vat is dangerously close to empty, but – at that very instant – it is also filling back up at a torrential rate of 2 cubic meters per minute. All praise to Ninkasi!

Whatever the graph of $V(t)$ may look like overall, we know it must pass through $(6, 0.0001)$, dropping almost down to the t-axis (which would signify an empty vat), but at that very point, we know the graph must also exhibit a strong sign of recovery: an upward-thrusting slope of 2. ♦

* *Squared seconds* are not physically meaningful units, hence the parenthetical sigh. One must, like Leopold Bloom, remember what m/s^2 actually *means*: "Thirtytwo feet per second, per second. Law of falling bodies: per second, per second. They all fall to the ground... Per second, per second. Per second for every second it means." (James Joyce, *Ulysses*)

Exercises.

6. Judging by the graph at right…

a) What is the approximate numerical value of $f'(3)$?

b) Arrange in numerical order: $f'(1),\ f'(2),\ f'(3),\ f'(4),\ f'(6)$.

c) How many solutions does $f(x) = 0$ have in the interval $[2,8]$?

d) How many solutions does $f'(x) = 0$ have in the interval $[2,8]$?

e) Which *integers* in $[2,8]$ satisfy the inequality $f(x)f'(x) < 0$?

f) True or false: $f'(\pi) > 0$. g) True or false: $[f(5)]/3 > f'(7)$.

7. True or false: If $f(x) = x^2$, $g(x) = \ln x$, and $h(x) = 1/x$, then…

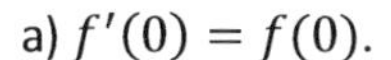

a) $f'(0) = f(0)$. b) $g(x) > 0$ for all x in the domain of g.

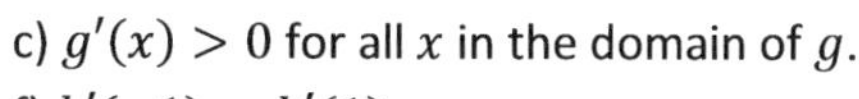

c) $g'(x) > 0$ for all x in the domain of g.

d) $g(1) = g'(1)$. e) $h'(x) < 0$ for all x in the domain of h. f) $h'(-1) = h'(1)$.

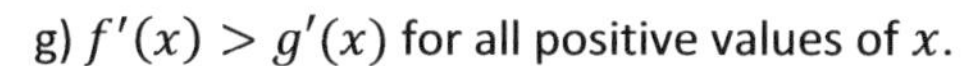

g) $f'(x) > g'(x)$ for all positive values of x.

8. Given the functions in the previous problem, arrange the following in numerical order: $f'(5),\ g'(5),\ h'(5)$.

9. Consider the constant function $f(x) = 5$. What is $f'(0)$? What is $f'(5)$? What is $f'(x)$? What can be said about the derivative of a constant function in general?

10. Consider the general linear function $g(x) = ax + b$. What is $g'(x)$?

11. The function $h(x) = |x|$ is defined for all real values of x, but $h'(x)$ has a slightly smaller domain. What is it? [*Hint: See exercise* 5.] Also, thinking geometrically, write down a formula for $h'(x)$.

12. Let $f(x) = \sqrt{4 - x^2}$.

a) Sketch graphs of f and f' on the same set of axes.

[*Hint: If you don't know the graph of f, square both sides of $y = \sqrt{4 - x^2}$; you'll know **that** equation's graph. Then observe that for each x-value on this "squared" graph, there are two y-values: one positive, one negative. Hence, if we solve its equation for its positive y-values (to recover the strictly positive function $y = \sqrt{4 - x^2}$), half of the graph will disappear; what remains is the graph of f. With that in hand, you can sketch the graph of f' by staring at the graph of f and thinking broadly about how its slope changes as x varies in its domain.*]

Find the exact values of the following.

b) $f'(0)$ c) $f'(1)$ d) $f'(\sqrt{2})$ e) $f'(\sqrt{3})$

[*Hint: Think about exercise* 2, *and recall how perpendicular lines' slopes are related.*]

13. Galileo fires a physics textbook out of a cannon. After t seconds, its height will be $h = -4.9t^2 + v_0 t$, where v_0 represents the book's initial upwards velocity in meters per second. Obviously, the book attains each height in its range (apart from its maximum height) at two separate moments: once going up, once coming down.

Remarkably, the book's speed will be the same at both moments. Without solving equations, explain why. [*Hint: Think geometrically. What does the height function's graph look like? How is speed encoded in it?*]

14. Rube Waddell throws a baseball at the full moon. Let $p(t)$ be the ball's height (in feet) t seconds after it leaves his hand. In terms of physics…

a) What does the quantity $p'(2)$ represent? b) What does the solution to the equation $p'(t) = 0$ represent?

c) What is the formula for $p''(t)$, and why is this so? d) If $p'(a) < 0$, then what is happening at time a?

15. Buffalo Bill goes ice skating. More particularly, he skates along a narrow frozen river running East/West, often reversing his direction by executing beautifully precise $180°$ turns. If we take a cigar that he dropped on the ice to be the origin, and we let East be the positive direction, then his position after t minutes can be described by the function $f(t)$, where distances are measured in meters. In physical terms...

a) What does $f'(t)$ represent? b) What does $|f'(t)|$ represent? c) What does $|f(t)|$ represent?
d) Could there be a time a at which $f(a) > 0$, but $f'(a) < 0$? Explain.
e) If $f''(t) = 0$ for all t in some interval (b, c), what is happening between $t = b$ and $t = c$?
f) If $f'(t) = 0$ for all t in some interval (d, e), what is happening between $t = d$ and $t = e$?
g) Suppose that $f'(t) < 0$ for all t in some interval (m, n), that $f'(n) = 0$, and that $f'(t) > 0$ for all t in some interval (n, p). What happened when $t = n$?

16. Physicists call the rate of change of acceleration *the jerk*. (Thus, the jerk is position's *third* derivative.)
a) If distance is measured in meters, and time in seconds, what are the units of the jerk?
b) What, if anything, can be said about a freely-falling object's jerk?

17. Suppose that the function $T(h)$ gives the temperature on 11/1/2015 at 10:46 am (the time at which I'm typing these words) h meters above my house in Olympia, WA. Suppose further that the function $T'(h)$ is strictly negative (meaning that its value is negative for all heights h). What does this strictly negative derivative signify physically?

18. Buzz Aldrin is walking clockwise around the rim of a perfectly circular crater on the moon, whose radius is 2 miles. Let $s(t)$ be the distance (in miles, as measured along the crater's rim) he has walked after t minutes.

a) If $s'(t) = \pi/40$ for all t during his first lap, then how long will it take him to complete one lap?
b) If $s'(t) > 0$ and $s''(t) < 0$ throughout his second lap, will Buzz be walking faster when he begins his second lap or when he ends it?

19. A mysterious blob from outer space has volume $V(t)$ after t hours on Earth, where V is measured in cubic feet.
a) What does $V'(5)$ represent physically, and in what units is it measured?
b) If the graph of $V(t)$ shows that V is decreasing between $t = 24$ and $t = 48$, what do we know about $V'(t)$?

A Gift From Leibniz: d-Notation

Now that you know what a derivative is, you are ready for the special notation introduced by Gottfried Wilhelm Leibniz, one of the intellectual giants in the story of calculus.

You are no doubt familiar with the "delta notation" commonly used to describe the *change* in a variable.* Recall in particular that we find a line's slope as follows: From the coordinates of any two of its points, we compute the "rise" (Δy) and "run" (Δx); the slope is then "the rise over the run", $\Delta y/\Delta x$.

In calculus, we use Leibniz's analogous "d-notation" to describe *infinitesimal* change. If m represents a magnitude (length, temperature, or what have you), then the symbol **dm represents an *infinitesimal change* in m.**

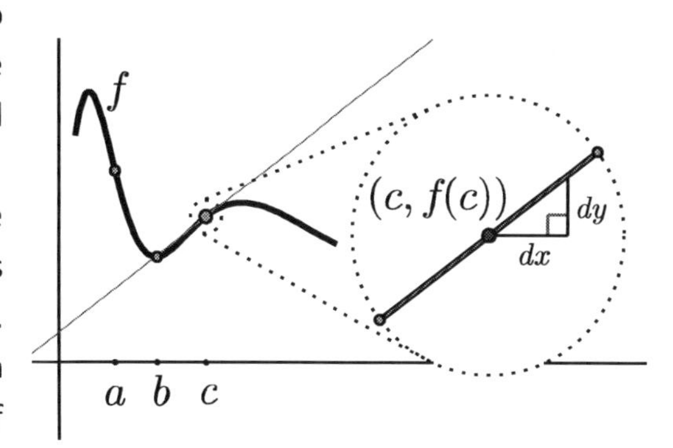

Since a curve is straight on an infinitesimal scale, the slope of its tangent (i.e. its derivative) at any given point is given by the ratio of infinitesimals $\boldsymbol{dy/dx}$, as in the figure. Accordingly, if $y = f(x)$, we frequently use "dy/dx" as an alternate notation for the derivative; that is, in place of $f'(x)$, we often write dy/dx.

The two derivative notations peacefully coexist. We use whichever one is more convenient in a given situation.

Prime notation's prime advantage is that it provides a clear syntactic place for the derivative's input. For example, consider the figure above. When the input is c, the derivative of f is approximately $3/4$. Using prime notation, we can express this fact quite compactly: $f'(c) \approx 3/4$. In contrast, expressing this same fact in **Leibniz notation** (as his d-notation is now called) is a bit more bothersome, as we must also supply a brief explanatory phrase:

$$\frac{dy}{dx} \approx 3/4 \text{ when } x = c.$$

The concluding phrase is essential, since dy/dx's values vary.†

Leibniz notation's many advantages will become ever clearer as you learn more and more calculus. For now, you should just appreciate how dy/dx encapsulates the derivative's *meaning*. When we see a derivative represented symbolically as dy/dx (an infinitesimal rise over an infinitesimal run), we are reminded that a derivative is a slope, a rate of change. In contrast, prime notation is totally arbitrary. After all, why use a prime and not, say, a dot?‡

Leibniz understood the importance of mental ergonomics. When he designed notation, he tried to fit it to the mind's natural contours. As we proceed through the course, we'll encounter numerous instances in which his notation will seem to do half of our thinking for us. Be thankful for Leibniz's gift.

---

* Examples: If h, the height of a sunflower (in feet), changes from 3 to 7 over a period of time, then $\Delta h = 4$ feet.
If the temperature T drops from $9°$ to $-11°$, then $\Delta T = -20°$.

† The notation $\left.\frac{dy}{dx}\right|_c$ is occasionally used for the derivative's value at c, but is very awkward. I won't use it in this book.

‡ Newton, in fact, did use a dot for derivatives. Newton and Leibniz are usually credited as the independent co-creators of calculus, which oversimplifies history a great deal, but is still a good first approximation to the truth. A bitter priority dispute arose between these two men and their followers.

Exercises.

20. If $y = f(x)$, rewrite the following statements in Leibniz notation: $f'(2) = 3,\ f'(\pi) = e$.

21. If $y = f(x)$, rewrite these statements in prime notation: $\frac{dy}{dx} = 1$ when $x = -2$, $\frac{dy}{dx} = \ln 5$ when $x = \sqrt{5}$.

22. If $f(2) = 6$ and $f'(2) = 3$ for a function f, then in an infinitesimal neighborhood around $x = 2$, the function's output increases by 3 units for each 1 unit of increase in the input. (Be sure you see why: Think geometrically.)

For most functions, the preceding statement is still approximately true for small *real* (i.e. not infinitesimal) neighborhoods of $x = 2$. Thus, increasing the input variable from 2 to 2.001 should cause the output variable to increase from 6 to approximately $6 + 3(.001) = 6.003$.

a) Convince yourself that the approximation described in the preceding paragraph should indeed be reasonable (even though the neighborhood isn't infinitesimal), provided the graph of f isn't too intensely curved near the point $(2, 6)$. As always, draw pictures to aid your intuition.

b) If $f(5) = 8$, and $f'(5) = 2.2$, approximate the value of $f(5.002)$.

c) If $g(0) = 2$, and $g'(0) = -1.5$, approximate the value of $g(0.003)$.

d) If $s(x) = \sin x$ (where x is measured in radians), we'll soon be able to prove that $s'(\pi) = -1$. Assuming this fact for now, approximate the value of $\sin(3.19)$. Then check your answer with a calculator.

e) TRUE of FALSE (explain your answer): If h is a function such that $h(3) = 2$ and $h'(3) = 4$, it is reasonable for us to assume that $h(10) \approx 2 + 4(7) = 30$.

23. We often read dy/dx aloud as "the derivative of y *with respect to* x". Similarly, dz/dq is the derivative of z with respect to q. In applications, this feature of Leibniz notation helps us keep track of units of measurement. For example, if K represents an object's kinetic energy (in joules) at time t (in seconds), then dK/dt is the derivative of kinetic energy *with respect to time*; it tells us how kinetic energy changes *in response to changes in time*. Moreover, the Leibniz notation makes it clear that dK/dt is measured in joules per second.

a) If, in the preceding kinetic energy example, we know that when $t = 60$, we have $K = 80$ and $dK/dt = 12$, then roughly how much kinetic energy might we reasonably expect the object to have when $t = 60.5$?

b) Suppose that A represents an area that grows and shrinks over time. Use Leibniz notation to express the following: After 5 minutes, the area is shrinking at a rate of 2 square meters per minute.

c) If $z = g(t)$, rewrite $g'(6) = 4$ in Leibniz notation.

d) Let C be the total cost (in dollars) of producing Q widgets per year. Boosting Q past certain values might require costly changes such as purchasing more machinery or hiring more employees. The derivative dC/dQ is known in economics as the "*marginal* cost function". In what units would values of dC/dQ be measured?

24. Leibniz notation's explicit reference to the derivative's input variable (see the previous exercise) is especially useful when "the" input variable can be viewed in multiple ways.

Consider, for instance, a conical vat. Suppose water pours into the (initially empty) vat at a constant rate of 3 ft^3/min. Let V be the volume of water in the cone, and let h be the water's "height", as indicated in the figure. Naturally, we can view V as a function of t, the time elapsed since the water began pouring in. However, we can also view V as a function of h; if the water's height is known, then the volume of the water is, in principle, determined – regardless of whether you know *how* to determine its numerical value.

Since V can be considered a function of t or h, we can distinguish between two different derivatives of V: dV/dt and dV/dh. The former measures the rate at which the volume changes with respect to *time*. This, we are told, is constant: $dV/dt = 3$ ft^3/min. The latter, dV/dh, measures the rate at which the volume changes with respect to the water's *height*. A little thought will convince you that this is *not* a constant rate of change.

a) Have the little thought mentioned in the previous sentence. Namely, to convince yourself that dV/dh is *not* a constant function of h, imagine two different situations corresponding to different values of h. First, let h be very small, so that there is hardly any water in the vat. If, in this case, we increase h by a tiny amount dh, consider the resulting change in volume, dV. *Draw a picture, and indicate what dV represents geometrically.* (You need not calculate anything.) Second, let h be relatively large, so that the vat is, say, $3/4$ full. Again, imagine increasing h by the same tiny amount dh. *Draw another picture and think about what dV represents geometrically.* Since applying the same little nudge to the input variable dh yields different changes to the output variable, dV, the ratio dV/dh has a different value in each case. Hence, dV/dh is not constant.

b) Which is greater: dV/dh when h is small, or dV/dh when h is large?

c) If the cone were upside down, so that water poured into its vertex at a constant rate, which would be greater: dV/dh when h is small, or dV/dh when h is large? *Draw a picture.*

d) If the water were pouring into a *spherical* tank of radius 10 feet, rank the following in numerical order: dV/dh when $h = 1$, dV/dh when $h = 5$, dV/dh when $h = 10$, dV/dh when $h = 17$.

e) Give an example of a shape for a tank that would ensure that dV/dh is constant if dV/dt is constant.

25. When $y = f(x)$, we can write dy/dx in the form $d(f(x))/dx$. Thus, for example, we may rewrite

$$\text{If } y = \tan x, \text{ then } \frac{dy}{dx} = 4 \text{ when } x = \frac{\pi}{3},$$

in the following more concise form:

$$\frac{d(\tan x)}{dx} = 4 \text{ when } x = \frac{\pi}{3}.$$

Naturally, all the usual interpretations hold. Here, for example, the notation is telling us that in the infinitesimal neighborhood of $x = \pi/3$, the output value of $\tan x$ increases by four units for each unit by which its input x is increased. Use this notation to rewrite the following statements.

a) If $y = \ln x$, then $\frac{dy}{dx} = 5$ when $x = \frac{1}{5}$.

b) If $y = 3x^3 + 1$, then $\frac{dy}{dx} = 36$ when $x = 2$.

c) If $y = 2^x$, then $\frac{dy}{dx} = \ln 16$ when $x = 2$.

d) If $y = -4x + 2$, then $\frac{dy}{dx} = -4$.

e) In part d, why wasn't it necessary to include a qualifying statement about an x-value? *Think geometrically.*

An Infinitesimal Bit of an Infinitesimal Bit

"So naturalists observe, a flea
Hath smaller fleas that on him prey;
And these have smaller fleas to bite 'em.
And so proceed *ad infinitum*."
- Jonathan Swift, "On Poetry: A Rhapsody"

Never forget: Ultimately, we are interested in *real*-scale phenomena. We work with infinitesimal bits of real magnitudes (dx, dz, or what have you) precisely because they help us understand real phenomena. In contrast, infinitesimal bits *of infinitesimals* ("second-order infinitesimals" such as $(dx)^2$ or $du \cdot dv$) mean nothing to us; when they appear in the same context as real magnitudes, we simply disregard them as though they were zeros. For instance, if we expand the binomial $(x + dx)^2$ to obtain

$$(x + dx)^2 = x^2 + 2x(dx) + (dx)^2,$$

we treat the second-order infinitesimal $(dx)^2$ as a zero, and thus we write $(x + dx)^2 = x^2 + 2x(dx)$.

It helps to imagine that the "calculus microscope" we described earlier (on the chapter's first page) can magnify first-order infinitesimals to visibility, but is too weak to detect higher-order infinitesimals. This "weakness" actually puts our eye in an ideal position – neither too close, nor too far away from the real magnitudes that we wish to describe. Could a more powerful microscope make sense of higher-order infinitesimals? Perhaps, but we need not concern ourselves with such questions here; we are interested in infinitesimals not for their own sake, but rather for what they tell us about ordinary, real-scale phenomena. For that purpose, first-order infinitesimals suffice.

Exercises.

26. Expand the following.

a) $(x - dx)^2$ b) $(x + dx)^2 - x^2$ c) $(x + dx)^3$ d) $(u + du)(v + dv)$

27. If we increase a function f's input from x to $x + dx$, its output changes from $f(x)$ to $f(x + dx)$. Consequently, the expression $\boldsymbol{d(f(x))}$, which denotes the infinitesimal change in f's value, is $\boldsymbol{f(x + dx) - f(x)}$.

[Thus, for example, $d(x^2) = (x + dx)^2 - x^2$. And so, by exercise $26b$, $d(x^2) = 2xdx$.]

Your problem: Show that…

a) $d(x^3) = 3x^2dx$ b) $d(3x^2) = 6xdx$ c) $d(ax^2 + bx + c) = (2ax + b)dx$.

28. Expand the following infinitesimal changes by writing them as *differences*.

a) $d(f(x))$ b) $d(5f(x))$ c) $d(f(x) + g(x))$ d) $d(f(x)g(x))$ e) $d\left(f(g(x))\right)$

29. On the graph of $y = x^2$, Let P be the point (3,9). Let Q be a point, infinitesimally close to P, whose coordinates are $(3 + dx,\ (3 + dx)^2)$. As we move from P to Q, the infinitesimal change in x is dx.

a) Express dy, the corresponding infinitesimal change in y, in terms of dx.

b) Since $y = x^2$, we have $dy = d(x^2)$. Use this, and your result from part (a) to compute $\frac{d(x^2)}{dx}$ when $x = 3$.

c) Use the ideas in this problem to find $\frac{d(x^2)}{dx}$ when $x = -1/2$.

30. a) Show that $\dfrac{\frac{u+du}{v+dv} - \frac{u}{v}}{dx} = \dfrac{vdu - udv}{v^2dx}$ b) Show that the right-hand side can be rewritten as $\dfrac{v\left(\frac{du}{dx}\right) - u\left(\frac{dv}{dx}\right)}{v^2}$.

The Derivative of $y = x^2$

Differential calculus teaches us nothing about linear functions; if we put the graph of a linear function under the calculus microscope, we see the same old straight line that had been visible to the naked eye. In exercise 10, you learned all there is to know about linear functions' derivatives: Linear functions have constant slopes, so their derivatives are constant functions. End of story.

Calculus exists to analyze *nonlinear* functions. Perhaps the simplest nonlinear function is $y = x^2$, whose derivative we'll now compute. This will be our first nontrivial example of a derivative. Note well: We are not merely looking for this function's derivative at a particular point (say, dy/dx when $x = 3$), which is a *number* (such as you found in exercise 29b); we seek the derivative dy/dx itself, a *function*. Once we've found a formula for dy/dx, we can evaluate it wherever we wish.

Problem. Find the derivative of the function $y = x^2$.

Solution. At each point x in the function's domain, the derivative's output will be the function's local rate of change there, dy/dx. To compute this ratio, we observe that when x is increased by an *infinitesimal* amount dx (which must be exaggerated in the figure!), the corresponding change in y is

$$\begin{aligned} dy &= (x + dx)^2 - x^2 \\ &= 2x(dx).^* \end{aligned}$$

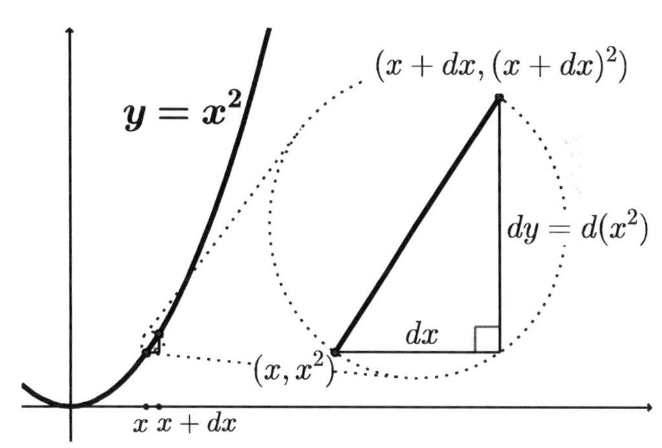

Thus, for any x in the domain, we have

$$\frac{dy}{dx} = \frac{2x(dx)}{dx} = \mathbf{2x}. \quad \blacklozenge$$

We've just proved that the derivative of x^2 is $2x$. That is,

$$\frac{d(x^2)}{dx} = 2x.$$

We often state results of this type ("the derivative of A is B") in the following alternate form:

$$\frac{d}{dx}(x^2) = 2x.$$

Here, we think of the symbol $\boldsymbol{d/dx}$ as an *operator* that turns a function into its derivative.†

Our first substantial project in differential calculus, which we'll begin after the next set of exercises, will be to discover a method for rapidly finding the derivative of any polynomial whatsoever.

---

* See exercise $26b$ above.

† Just as a function turns numbers into numbers, an *operator* turns functions into functions.

Exercises.

31. Knowing $\frac{d(x^2)}{dx} = 2x$ in general, you should need only seconds to find $\frac{d(x^2)}{dx}$ when $x = 4$ in particular. Find it.

32. Find the equation of the line tangent to the graph of $y = x^2$ at point $(-3,9)$.

33. a) Find the coordinates of the one point on the graph of $y = x^2$ at which the function's slope is exactly -5.

b) Find the equation of the tangent to the graph at the point you found in part (a).

34. Is there a non-horizontal tangent to $y = x^2$ that passes through $(1/3\,,0)$? If so, find the point of tangency.

35. Use the ideas in this section to find $\frac{d(ax^2+bx+c)}{dx}$.

36. a) Use the ideas in this section to find $\frac{d(x^3)}{dx}$. b) Use part (a) to find $\frac{d(x^3)}{dx}$ when $x = \sqrt{2}$, and when $x = \pi$.

c) Find the equation of the line tangent to the graph of $y = x^3$ at point $(1,1)$.

d) Does the tangent line from part (c) cross the graph again? If so, where? If not, how do you know?

[*Hint: At some point, you'll need to solve a cubic equation. Even though you (presumably) do not know how to solve cubics in general, you can solve this particular one because you already know one of its solutions.*]

37. The notations $\frac{d(f(x))}{dx}$ and $\frac{d}{dx}\big(f(x)\big)$ are equivalent and are used interchangeably. To accustom yourself to these notational dialects, rewrite the following expressions in the other form. (Yes, this exercise is trivial.)

a) $\frac{d(x^2)}{dx}$ b) $\frac{d(\sin x)}{dx}$ c) $\frac{d}{dx}(\ln x)$ d) $\frac{d}{dx}\left(\frac{1}{x}\right)$

Derivatives of Polynomials

'He is like a mere x. I do not mean x the kiss symbol but, as we allude in algebra terminology, to denote an unknown quantity.'
'What the hell this algebra's got to do with me, old feller?'

- *All About H. Hatterr*, G.V. Desani.

Let us revisit some basic algebra you learned on your mother's knee.

How does one multiply algebraic expressions such as $(a + b + c)(d + e + f)$? Well, each term in the first set of parentheses must "shake hands" with each term in the second set. The *sum* of all such "handshakes" (i.e. multiplications) is the product we seek.

For example, to multiply $(a + b + c)(d + e + f)$, first a shakes hands with each term in the second set (yielding ad, ae, and af), then b does (bd, be, and bf), and finally, c does (cd, ce, and cf). Thus,

$$(a + b + c)(d + e + f) = ad + ae + af + bd + be + bf + cd + ce + cf.$$

Naturally, this works regardless of how many terms are in each parenthetical expression. Applied, for instance, to the very simple product $(a + b)(c + d)$, the handshake game produces the familiar "FOIL" expansion you learned in your first algebra course.

If one wishes to multiply not two, but *three* expressions, then each "handshake" must be a *three-way* handshake, with one handshaker drawn from each of the three expressions. For example,

$$(a + b)(c + d)(e + f) = ace + acf + ade + adf + bce + bcf + bde + bdf.$$

Similarly, multiplying four expressions requires four-way handshakes, while multiplying n expressions, as we'll need to do in a moment, requires n-way handshakes. Let us rest our hands and return to calculus.

Problem. Find the derivative of the function $y = x^n$, where n is a whole number.

Solution. We begin by noting that

$$\frac{d(x^n)}{dx} = \frac{(x+dx)^n - x^n}{dx}.$$

To proceed, observe that the binomial in the numerator is

$$(x+dx)^n = \overbrace{(x+dx)(x+dx)\cdots(x+dx)}^{n \text{ times}}.$$

To multiply this out, we must sum all possible n-way "handshakes" of the sort described above.

The simplest of these n-way handshakes will involve all n of the x's and none of the dx's. This handshake's contribution to the expansion of $(x+dx)^n$ is obviously $\boldsymbol{x^n}$.

Among the many other n-way handshakes, some will involve $(n-1)$ of the x's and one dx. In fact, there will be exactly n handshakes of this sort. (The first of them involves the dx from the *first* parenthetical expression and the x's from all the other groups; the second involves the dx from the *second* parenthetical expression and the x's from all the other groups – and so forth.) Since each of these n handshakes will add an $x^{n-1}dx$ to the expansion of $(x+dx)^n$, their net contribution to the expansion will be $\boldsymbol{nx^{n-1}dx}$.

Each of the remaining handshakes involves at least two dx's, so their contributions to the expansion will be higher-order infinitesimals, which means that we can simply disregard them! Therefore, the expansion we seek is $(x+dx)^n = x^n + nx^{n-1}dx$.

Consequently,

$$\begin{aligned}\frac{d(x^n)}{dx} &= \frac{(x+dx)^n - x^n}{dx} \\ &= \frac{(x^n + nx^{n-1}dx) - x^n}{dx} \\ &= nx^{n-1}.\end{aligned}$$

♦

The result we've just established is important enough to merit its own box.

> **The Power Rule (Preliminary Version).**
>
> For all whole numbers n,
>
> $$\boldsymbol{\frac{d}{dx}(x^n) = nx^{n-1}}.$$

This is a preliminary version inasmuch as we will eventually prove that the power rule holds not just for whole number powers, but for all *real* powers whatsoever.

Example. By the power rule, $\frac{d}{dx}(x^7) = 7x^6$. ♦

A quick note on language: We do *not* say that we "derive" x^7 to obtain its derivative, $7x^6$. Rather, we "take the derivative of" x^7.* Many students commit this faux pas. Don't be one of them.

Over the next few chapters, you'll learn how to take many functions' derivatives. Along with derivatives of specific functions (x^n, $\sin x$, $\ln x$, etc.), you'll learn *structural* rules that let you take derivatives of nasty functions built up from simple ones (such as $x^2 \ln x + 4\sin(5x)$). The first structural rules we'll need are the derivative's *linearity properties*.

Linearity Properties of the Derivative.

i. The derivative of **a constant times a function** is the constant times the function's derivative:

$$\frac{d}{dx}\big(cf(x)\big) = c\frac{d}{dx}\big(f(x)\big).$$

ii. The derivative of **a sum of functions** is the sum of their derivatives:

$$\frac{d}{dx}\big(f(x) + g(x)\big) = \frac{d}{dx}\big(f(x)\big) + \frac{d}{dx}\big(g(x)\big).$$

Proof. To prove the first linearity property, we just follow our noses:

$$\frac{d}{dx}\big(cf(x)\big) = \frac{d\big(cf(x)\big)}{dx} = \frac{cf(x+dx) - cf(x)}{dx}$$
$$= c\left(\frac{f(x+dx) - f(x)}{dx}\right) = c\left(\frac{d\big(f(x)\big)}{dx}\right) = c\frac{d}{dx}\big(f(x)\big).$$

Proving the second property is just as simple.

$$\frac{d}{dx}\big(f(x) + g(x)\big) = \frac{d\big(f(x)+g(x)\big)}{dx} = \frac{\big(f(x+dx) + g(x+dx)\big) - \big(f(x) + g(x)\big)}{dx}$$
$$= \frac{\big(f(x+dx) - f(x)\big) + \big(g(x+dx) - g(x)\big)}{dx} = \frac{f(x+dx) - f(x)}{dx} + \frac{g(x+dx) - g(x)}{dx}$$
$$= \frac{d\big(f(x)\big)}{dx} + \frac{d\big(g(x)\big)}{dx} = \frac{d}{dx}\big(f(x)\big) + \frac{d}{dx}\big(g(x)\big). \quad \blacksquare$$

If that proof gave you any trouble, please work through it again after revisiting exercises 27, 28, and 37. The justification for each equals sign in the proof should be crystal clear to you.

Using the power rule and the linearity properties, we can find any polynomial's derivative in a matter of seconds, as the following example and exercises will convince you.

Example. Find the derivative of $y = 5x^6 + 3x^4$.

Solution. $\frac{d}{dx}(5x^6 + 3x^4) = \frac{d}{dx}(5x^6) + \frac{d}{dx}(3x^4)$ (by one of the linearity properties)

$= 5\frac{d}{dx}(x^6) + 3\frac{d}{dx}(x^4)$ (by the other linearity property)

$= 5(6x^5) + 3(4x^3)$ (by the power rule)

$= 30x^5 + 12x^3$. ♦

---

* One may also use the verb *differentiate* here. (E.g. If we differentiate x^7, we obtain $7x^6$.) However, since this sense of "differentiate" has nothing in common with the verb's ordinary meaning, those who use it must take care to differentiate differentiate from differentiate.

Exercises.

38. The power rule is geometrically obvious in the special cases when $n = 1$ or $n = 0$. Explain why.

39. If we combine the power rule and the first linearity property, we find that $\frac{d}{dx}(cx^n)$ is equal to what?

40. Use the result of exercise 39 to find the following derivatives in one step:

a) $\frac{d}{dx}(5x^3)$ b) $\frac{d}{dx}(-3x^7)$ c) $\frac{d}{dx}\left(\frac{\pi}{2}x^2\right)$

41. By combining the result of exercise 39 and the second linearity property, find the following derivatives.

a) $\frac{d}{dx}(3x^3 + 2x^2)$ b) $\frac{d}{dx}(-10x^5 + \frac{1}{4}x^3)$ c) $\frac{d}{dx}(\sqrt{2}x + \sqrt{3})$

42. Prove that the derivative of **a difference of functions** is the difference of their derivatives.

43. Use the result of the previous exercise to find the following derivatives.

a) $\frac{d}{dx}(5x^3 - 2x^4)$ b) $\frac{d}{dx}(3x^2 - 5)$ c) $\frac{d}{dx}\left(-\frac{2}{5}x^{10} - \pi\right)$

44. Convince yourself that the derivative of a sum of *three* (or more) functions is the sum of their derivatives. Then use this fact to compute the following, ideally writing down each derivative in a single step:

a) $\frac{d}{dx}(2x^3 + 4x^2 + 5x + 1)$ b) $\frac{d}{dx}\left(\frac{3}{4}x^3 - 9x^2 - \sqrt{5}x + 2\right)$ c) $\frac{d}{dx}(-x^6 + x^3 - 4x^2 + 3)$

45. The derivative of a product of functions is *not* the product of their derivatives! Show, for example, that

$$\frac{d}{dx}(x^5x^3) \neq \left(\frac{d}{dx}(x^5)\right)\left(\frac{d}{dx}(x^3)\right).$$

46. a) True or false: $\frac{d}{dx}((2x+1)^3) = 3(2x+1)^2$. Explain your answer.

b) What, in fact, is the derivative of $y = (2x+1)^3$?

47. The derivative of a quotient of functions is *not* the quotient of their derivatives! Demonstrate this by providing a counterexample, as in exercise 45.

48. If $y = 2\pi^3$, what is dy/dx?

49. What does the power rule tell us about $\frac{d}{dx}(3^x)$?

50. Using symbols other than x and y for a function's independent and dependent variables does not change the formal rules for finding derivatives. With this in mind, find the derivatives of these functions:

a) $f(t) = 2t^3 - 3t^2 + 5$ b) $g(z) = \frac{1}{4}z^4 + \frac{1}{3}z^3 + \frac{1}{2}z^2 + z$ c) $A(r) = \pi r^2$

51. Recall that the graph of any quadratic function (of the form $y = ax^2 + bx + c$) is a *parabola* whose axis of symmetry is parallel to the y-axis. Clearly, such a graph has a horizontal tangent only at its vertex. This observation yields a quick way to find the vertex's x-coordinate: Set the quadratic's derivative equal to zero. Be sure you understand this idea, then use it – together with the fact that a parabola of this sort opens up or down according to whether its leading coefficient is positive or negative – to sketch graphs of the following quadratics. Include the coordinates of each parabola's vertex and of any intersections with the axes.

a) $y = x^2 + 3x + 4$ b) $f(x) = -2x^2 + 3x - 4$ c) $g(x) = \pi x^2 + ex + \sqrt{2}$

52. Find the equation of the line tangent to $y = x^3 - 2x^2 + 3x + 1$ at $(1,3)$.

53. There is exactly one tangent to $y = x^3$ that passes through $(0,2)$. Find the point of tangency.

One Last Example

"Imagine, if you will, that the stone, while in motion, could think... Such a stone, being conscious merely of its own endeavor... would consider itself completely free, would think it continued in motion solely by its own wish. This then is that human freedom which all men boast of possessing, and which consists solely in this: that men are conscious of their own desire, but ignorant of the causes whereby that desire has been determined."

- Spinoza, in a letter to G.H. Schuller (October, 1674)

Your new ability to take polynomials' derivatives lets you solve otherwise tricky applied problems.

Example. Spinoza stands at a cliff's edge, 100 feet above the ocean, and hurls a stone. Its height (relative to the ocean) after t seconds is given by the formula $s(t) = -16t^2 + 64t + 100$. Answer the following questions.

a) Find the stone's vertical velocity when $t = 0.5$ and when $t = 2.5$.
b) What is the stone's maximum height?
c) How fast is the stone moving downwards at the instant when it hits the water?

Solution. Since $s(t)$ gives the stone's vertical position (height in feet) after t seconds, its derivative, $s'(t) = -32t + 64$, gives the stone's vertical *velocity* (in ft/sec) after t seconds.

Thus, half a second after leaving Spinoza's hand, the stone's vertical velocity is $s'(0.5) = 48$. That is, at that particular instant, it is moving *upwards* at 48 ft/sec. After 2.5 seconds, its vertical velocity is $s'(2.5) = -16$. Hence, at *that* instant, the stone is moving *downwards* at 16 ft/sec.

Clearly, the stone will rise for a while (have positive vertical velocity), then fall (have negative vertical velocity). The stone will reach its maximum height at the instant when it has stopped rising, but has not yet begun to fall. This occurs when its vertical velocity is *zero*. Solving $s'(t) = 0$, we find that this maximum height occurs at $t = 2$. Consequently, the stone's maximum height will be $s(2) = 164$ feet above the ocean.

The stone hits the water when $s(t) = 0$. This is a quadratic equation; substituting its sole positive solution, $t = 2 + \sqrt{41}/2$ into our velocity function, we find that the stone's vertical velocity upon impact is $s'\left(2 + \sqrt{41}/2\right) \approx -102.4$ ft/sec. ♦

As discussed earlier, velocity's rate of change is *acceleration*. In the preceding example, the stone's position was given by $s = -16t^2 + 64t + 100$, from which we deduced its velocity function:

$$v = \frac{ds}{dt} = -32t + 64\,.$$

By taking the derivative of the velocity function, we can now determine the rock's acceleration function:

$$a = \frac{dv}{dt} = -32\,,$$

which agrees with Galileo's famous discovery: Any object in free fall (i.e. with no force other than gravity acting upon it) accelerates downwards at a constant rate of 32 feet per second per second.

Exercises.

54. Suppose a point is moving along a horizontal line. Define *right* as the positive direction. If the point's position (relative to some fixed origin) after t seconds is given by $s = -5 + 4t - 3t^2$, find the time(s) at which the point is momentarily at rest (i.e. when its velocity is zero), the times when the point is moving to the right, and the times when it is moving to the left.

55. Suppose two points are moving on the line from the previous problem, and their positions are given by

$$s_1(t) = t^2 - 6t \quad \text{and} \quad s_2(t) = -2t^2 + 5t.$$

a) Which point is initially moving faster?

b) When, if ever, will the two points have the same velocity?

c) When the clock starts, the two points occupy the same position, but they separate immediately thereafter. Where and when will they next coincide? What will their velocities be then? Will they meet a third time?

56. Molly Bloom throws an object down from the Rock of Gibraltar in such a manner that the distance (in meters) it has fallen after t seconds is given by the function $s = 30t + 4.9t^2$. How fast is the object moving downwards after 5 seconds? What is the object's acceleration then (in m/s^2)?

57. In exercise 50c, you showed that the derivative of a circle's area (with respect to its radius) is its circumference. Is this just a curious coincidence, or is there a deeper reason for it? Thinking geometrically will help you understand.

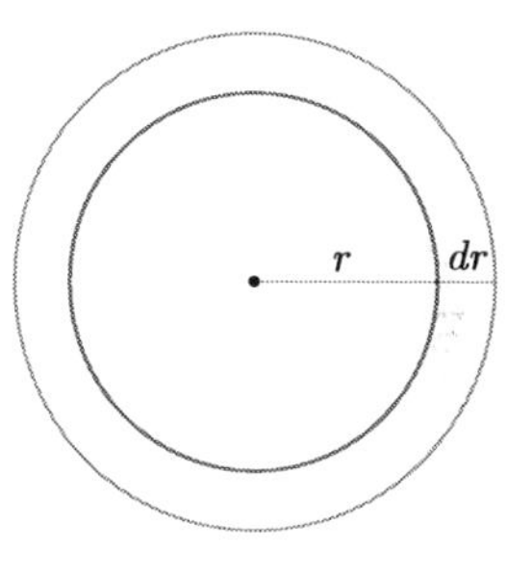

If a circle's radius r increases infinitesimally by dr (necessarily exaggerated in the figure), then its area A will increase infinitesimally by dA. In the figure, dA is the area of the infinitesimally thin ring bounded by the two circles. In this exercise, you'll consider two different geometric explanations of *why* dA/dr equals C, the original circle's circumference.

a) The first explanation is basically computational: Given that the outer circle's radius is $(r + dr)$, express dA, the area in the ring, in terms of r. Then divide by dr, and verify that the derivative dA/dr is indeed $2\pi r$.

b) The second explanation cuts right to the geometric heart of the phenomenon; it doesn't even involve the circle's area and circumference formulas.

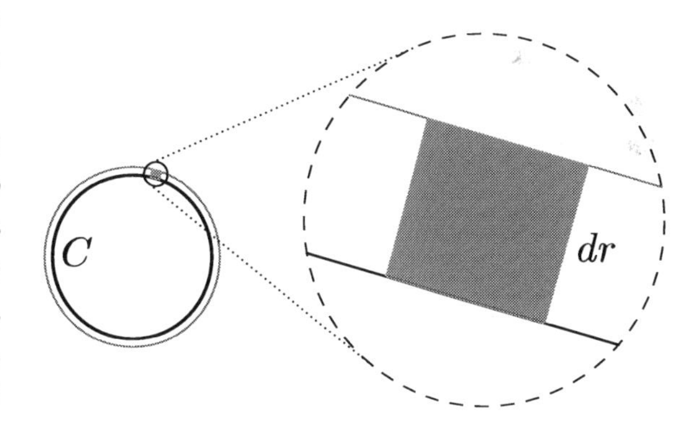

Consider the figure at right. The infinitesimally thin ring (whose area is dA) can be broken up into infinitesimal rectangles. Placing all the rectangles end to end, we can construct *one* rectangle, whose height and length will be dr and C respectively, and whose area must therefore be Cdr. However, its area must also be dA (since it was reformed from pieces of the ring).

It follows that $dA = Cdr$, which is equivalent to $dA/dr = C$, as claimed.

For many people, this argument provides a flash of insight that renders the formula $dA/dr = C$ obvious. Others, however, are uncomfortable with it, and wonder if it sweeps something important under the rug. Remarkably, both views can coexist in the same mind; one can feel the flash of geometric illumination, and yet still wonder if the means by which it was conveyed are entirely sound.

So, dear reader, did this argument help *you* see why the derivative of a circle's area (with respect to its radius) is its circumference? And is there any of part of the argument that particularly troubles you?

58. Recall that a sphere of radius r has volume $V = (4/3)\pi r^3$. The power rule tells us that $dV/dr = 4\pi r^2$, which is – as the previous problem might lead you to expect – the sphere's surface area. Using analogs of either (or ideally both) of the arguments in the previous problem, try to gain insight into *why* this fact must be true.

59. A square of side length x has area $A = x^2$ and perimeter $P = 4x$. Contrary to what one might expect from the previous two problems, the power rule shows that $dA/dx \neq P$. To understand this geometrically, consider the figure, remembering once again that as with all such schematic depictions of infinitesimals, you must imagine the dx's as being incomparably tinier than they appear.

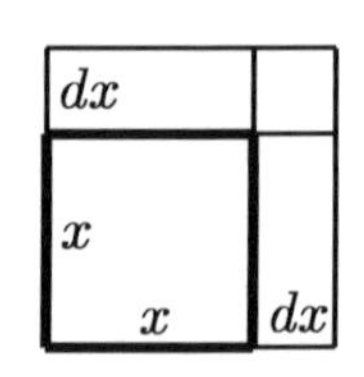

a) If A represents the original square's area (before the infinitesimal change to x), then to what part of the figure does dA correspond?

b) Use analogs of either (or both) of the arguments from exercise 57 to try to gain insight into why $dA/dx \neq P$.

60. Dippy Dan doesn't know how to communicate mathematical ideas. Collected below are eight samples of his garbled writing. Explain *why* his statements are gibberish, and suggest sensible alternatives that accurately convey what he is presumably trying to express.

a) $\frac{dy}{dx}(x^3) = 3x^2$.
b) If $y = 2x^4$, then $\frac{d}{dx} = 8x^3$.
c) $\frac{d}{dx}(\pi r^2) = 2\pi r$
d) If $y = -7x$, then $dy = -7$.
e) $x^8 = 8x^7$.
f) $2x^2 + 5x^2 \to 7x^2$.
g) $\frac{d}{dx}(3x^4) = 4(3x^3) \to 12x^3$.
h) $d(3x^2) = 6x$.
i) $x^2 \frac{d}{dx} = 2x$.

Chapter 2

The Differential Calculus Proper

The Derivatives of Sine and Cosine

To discover a function's derivative, we change its input infinitesimally, then find the resulting infinitesimal change in output, and finally, we take the ratio of these two infinitesimal changes. Let us do this for sine.

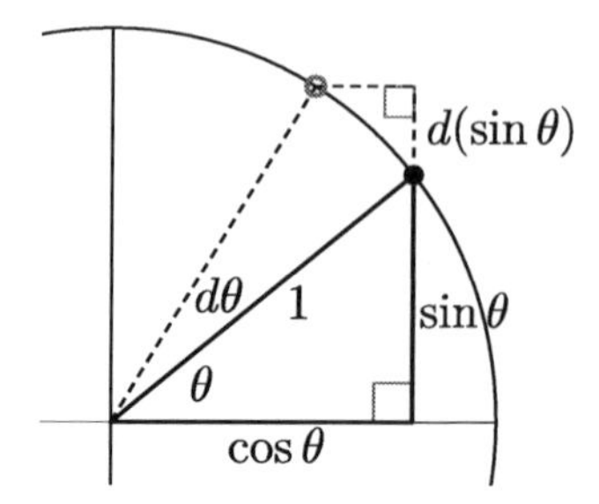

Consider the figure at right, which takes place on the unit circle. By the definition of sine, the sine of θ is the solid point's y-coordinate. Now we'll increase sine's input infinitesimally by $d\theta$ (from θ to $\theta + d\theta$). When we do so, sine's new output will be the *hollow* point's y-coordinate. Thus, the infinitesimal *change* in sine's output, $d(\sin\theta)$, is indeed the length of the segment that I've labelled as such in the figure.

Next, note that *if we measure angles in radians*, the length of the arc between the solid and hollow points must be $d\theta$.*

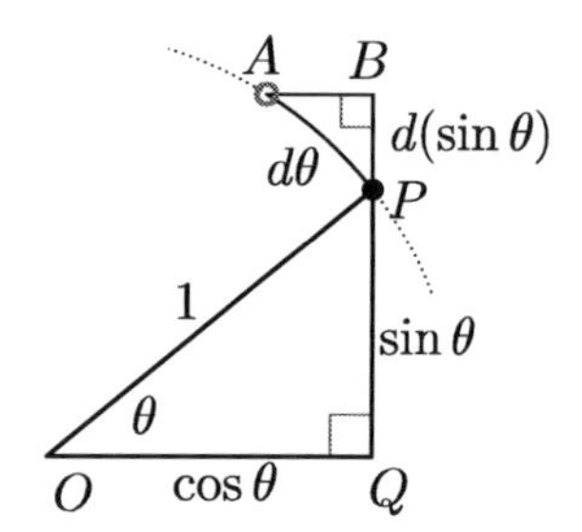

This gives us the picture at right. Note that "arc" AP is *straight*, since $d\theta$ is infinitesimal. Moreover, the infinitesimal right triangle of which it is part is similar to ΔOPQ. [Proof: Any circle is perpendicular to its radii, so $A\hat{P}O$ is a right angle. Thus, $B\hat{P}A$ is $O\hat{P}Q$'s complement, which – as a glance at ΔOPQ shows – is θ. Hence, $\Delta OPQ \sim \Delta PAB$ by AA-similarity.] Consequently, the ratio (leg-adjacent-to-θ):(hypotenuse) is the same in each triangle. That is,

$$\frac{d(\sin\theta)}{d\theta} = \frac{\cos\theta}{1}.$$

Thus we have the derivative we seek: **The derivative of $\sin\theta$ is $\cos\theta$.**

You'll be pleased to know that this intricate argument gives us cosine's derivative as a free bonus. We need only observe in the figure that $AB = -d(\cos\theta)$.† Ratios of corresponding parts being equal in similar triangles, we have

$$\frac{-d(\cos\theta)}{d\theta} = \frac{\sin\theta}{1}.$$

Multiplying both sides by -1 reveals that **the derivative of $\cos\theta$ is $-\sin\theta$.**

To sum up, we have proved the following important results. Mark the introductory phrase well!

> When θ is measured in radians,
>
> $\frac{d}{d\theta}(\sin\theta) = \cos\theta$, and $\frac{d}{d\theta}(\cos\theta) = -\sin\theta$.

* **Proof**: There are 2π radians in a full rotation, so an angle of $d\theta$ radians is $d\theta/2\pi$ of a full rotation. Accordingly, a central angle of $d\theta$ radians is subtended by $d\theta/2\pi$ of the circle's circumference. Since the unit circle's circumference is 2π, the arc length between the solid and hollow points is $d\theta/2\pi$ of 2π, which is indeed $d\theta$, as claimed.

† An explanation for the negative: Thinking about cosine's definition reveals that the infinitesimal change $d(\cos\theta)$ and the length AB have the same *magnitude*. Their *signs*, however, are opposite: AB, being a length, is necessarily positive, but since the figure shows that increasing θ by $d\theta$ causes a *decrease* in cosine, $d(\cos\theta)$ must be negative. Hence, $AB = -d(\cos\theta)$.

Exercises.

1. The variables in a function's formula are just placeholders. For instance, $f(x) = x^2$, $y = t^2$, and $h(z) = z^2$ all represent the same underlying *squaring function*, so they all have the same derivative (the *doubling function*), even if we express this fact with different symbols in each case: $f'(x) = 2x$, $dy/dt = 2t$, and $h'(z) = 2z$. This being so, write down the derivatives of the following functions, using notation suitable for each case.

a) $f(\theta) = \sin\theta$ b) $z = \cos t$ c) $x = \sin y$ d) $B = \cos\alpha$ e) $g(w) = \sin w$.

2. Combining this section's results with the derivative's linearity properties, find the derivatives of the following.

a) $f(x) = 3\sin x - 5\cos x$ b) $y = \pi t^2 + \sqrt{2}\sin t + e^5$ c) $g(x) = -\cos x - \sin x$

3. If $y = \cos x$, what is $(y')^2 + y^2$?

4. Pinpoint the precise place in our derivation of sine's derivative where we used radians. Then, by making appropriate adjustments in the rest of the argument, determine the derivative of $\sin\theta$ if one measures angles in *degrees*. You'll find the final result is slightly different (and messier) than when we measure angles in radians.

If calculus did not exist, neither would radians. Mathematicians and scientists adopted radian measure primarily to ensure that sine's derivative is as simple as possible. Any other angle measure (degrees, gradians, or what have you) yields a derivative to which an ugly constant clings like a barnacle.

5. Strictly speaking, the figure and argument in our derivation of sine's derivative covers only the case in which θ lies in the first quadrant. We can easily extend the result to all values of θ, though the details are a bit tedious, which is why I omitted them. If θ lies in the second quadrant, for example, increasing θ causes $\sin\theta$ to *decrease*, with the result that, in the figure, one of the infinitesimal right triangle's sides would be $-d(\sin\theta)$ instead of $d(\sin\theta)$. Draw a picture of this second-quadrant case, and verify that a compensating change in the *other* right triangle ensures that sine's derivative is still cosine. (Then, should you feel so inclined, you can cross the last "t" and dot the final "i" by verifying that the result still holds when θ lies in quadrants three or four.)

6. For very small values of x (where x is measured in radians), $\sin x \approx x$. Explain *why* this approximation, which is frequently used by scientists and engineers, holds. [*Hint: Find the tangent to the graph of* $y = \sin x$ *at* $x = 0$.]

7. To find the derivative of a composite function such as $y = \sin(5x + 2)$ requires a little trickery. The trick is to rewrite it in terms of its simple components:

$$y = \sin u, \quad \text{where} \quad u = 5x + 2.$$

So far, we've expressed y as a simple function of u, which in turn is a simple function of x. Having accomplished this, it is easy to find dy/dx. We need only make the general observation that

$$\frac{dy}{dx} = \frac{dy}{du} \cdot \frac{du}{dx}.$$

Applied to our particular function, this yields

$$\frac{dy}{dx} = (\cos u)5.$$

Substituting $5x + 2$ back in for u (and moving that 5 to its customary location), we have our derivative:

$$\frac{dy}{dx} = 5\cos(5x + 2).$$

Use this trick to find the derivatives of the following functions.

a) $y = \sin(3x + 6)$ b) $y = \cos(3x^2)$ c) $y = \sin(\cos x)$ d) $y = (3x + 1)^{50}$
e) $y = \sin^2 x$ f) $y = 2\sin x\cos x$ [*Hint: A trigonometric identity will help.*]

8. Everyone knows that in the figure at right, $OA = \cos\theta$ and $AB = \sin\theta$. Surprisingly few students (or even teachers) of trigonometry know that the *tangent* function also lives on the unit circle. Its location relative to the unit circle explains why the tangent function is called the tangent function.

a) Using similar triangles and the "SOH CAH TOA" definitions from right-angle trigonometry, prove that $PT = \tan\theta$.

b) The secant function also lives on the unit circle. Prove that $OP = \sec\theta$.

c) The Latin verbs *tangere* and *secare* mean "to touch" and "to cut" respectively. What does this have to do with lines PT, PO, and the circle? (A-ha: Now you know why secant is called secant.)

d) The one trigonometric identity everyone remembers is the "Pythagorean Identity," $\cos^2\theta + \sin^2\theta = 1$. Explain this famous identity's name by thinking about ΔOAB in the figure.

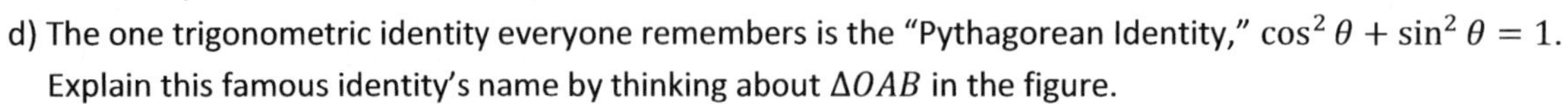

e) There is an alternate version of the Pythagorean identity that often comes in handy in integral calculus: $1 + \tan^2\theta = \sec^2\theta$. Explain how the truth of this identity, too, can be seen in the figure.

f) Those of a more algebraic mindset can derive the alternate Pythagorean identity from the ordinary one by dividing both sides by $\cos^2\theta$. Verify that this is so.

9. Now that you know where $\tan\theta$ and $\sec\theta$ live on the unit circle, we can find their derivatives. (We'll soon learn how to find them algebraically, but doing it geometrically is more aesthetically satisfying.)

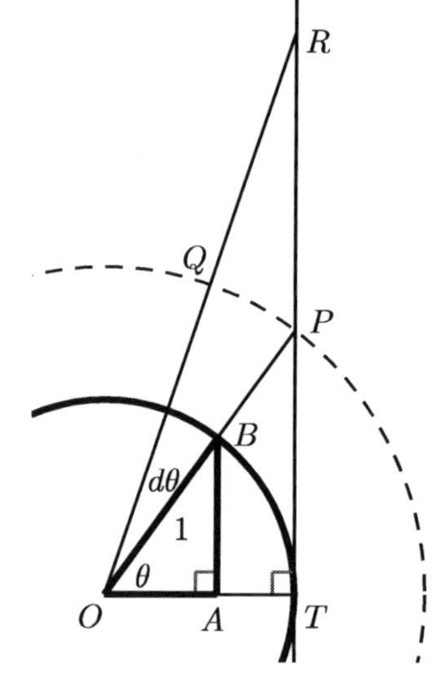

a) In the figure at right, the infinitesimal increment $d\theta$ is necessarily exaggerated. Convince yourself that, despite appearances, ΔPQR represents an infinitesimal *right* triangle.

b) Explain why $PR = d(\tan\theta)$.

c) *If we measure angles in radians*, explain why arc QP has length $\sec\theta\, d\theta$.

d) Explain why ΔPQR is similar to ΔOAB.

e) Use this similarity (plus a little algebra) to prove that

$$\frac{d(\tan\theta)}{d\theta} = \sec^2\theta.$$

f) Explain why $QR = d(\sec\theta)$.

g) Use similar triangles (plus a little algebra) to prove that

$$\frac{d(\sec\theta)}{d\theta} = \sec\theta\tan\theta.$$

10. Poor cotangent and cosecant, the least loved of the six trigonometric ratios, are also denizens of the unit circle.

a) In the figure at right, prove that $PC = \cot\theta$.

b) Prove that $PO = \csc\theta$.

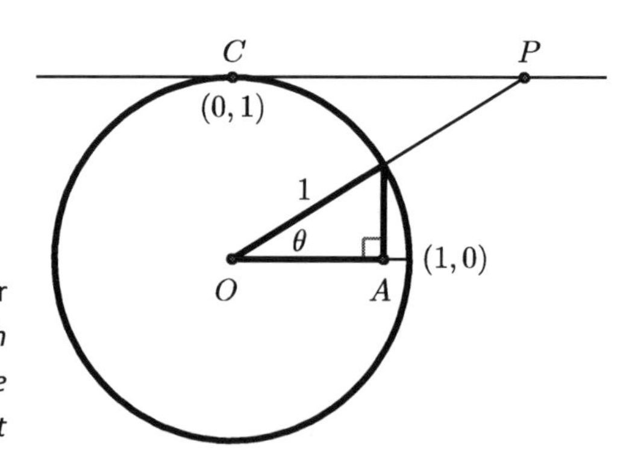

c) Use the figure to explain another alternate version of the Pythagorean identity: $1 + \cot^2\theta = \csc^2\theta$.

d) Explain how to derive the identity in part (c) from the ordinary Pythagorean identity with a little algebra.

e) Using arguments like those you used in exercise 9, discover the derivatives of $\cot\theta$ and $\csc\theta$. [*Hint: You'll need to watch out for negatives, as in the derivation of cosine's derivative above. For instance, you'll need to identify a line segment whose length is* $-d(\cot\theta)$.]

The Product Rule

The derivative of a product is *not* a product of derivatives. (You proved this in exercise 45 of Chapter 1.) Here is the actual rule for a product's derivative.

> **Product Rule.** If u and v are functions of x, then
> $$\frac{d}{dx}(uv) = u\frac{dv}{dx} + v\frac{du}{dx}.$$

Proof. If we increase the input x by an infinitesimal amount dx, then the outputs of the individual functions u and v change by infinitesimal amounts du and dv. Consequently, the value of their *product* changes from uv to $(u + du)(v + dv)$. With this in mind, we see that

$$\frac{d(uv)}{dx} = \frac{(u + du)(v + dv) - uv}{dx} = \frac{udv + vdu}{dx} = u\frac{dv}{dx} + v\frac{du}{dx}. \quad \blacksquare$$

The product rule is easy to use, as the following examples demonstrate.

Example 1. Find the derivative of $y = x^2 \sin x$.
Solution. Because we can view this function as a product uv, where $u = x^2$ and $v = \sin x$, we can apply the product rule. It tells us that

$$\begin{aligned}\frac{dy}{dx} &= (x^2)\frac{d}{dx}(\sin x) + (\sin x)\frac{d}{dx}(x^2) \\ &= x^2 \cos x + 2x \sin x. \quad \blacklozenge\end{aligned}$$

Example 2. Find $\frac{d}{dx}\big((3x^3 + x)\cos x\big)$.
Solution. By the product rule,

$$\begin{aligned}\frac{d}{dx}\big((3x^3 + x)\cos x\big) &= (3x^3 + x)\frac{d}{dx}(\cos x) + (\cos x)\frac{d}{dx}(3x^3 + x) \\ &= -(3x^3 + x)\sin x + (9x^2 + 1)\cos x. \quad \blacklozenge\end{aligned}$$

That's all there is to it.

Exercises.

11. Find the derivatives of the following functions.
a) $y = -3x^8 \cos x$ b) $y = \sin x \cos x$ c) $y = (9x^4 - x^3 + \pi^2)\sin x$ d) $y = \pi x^2(\sin x + \cos x)$

12. Compute the derivative of $y = (2x^2 + 3x)(5x^2 + 1)$ two different ways, and verify that the results are equal.

13. Derive a "triple product rule" for the derivative of uvw (where u, v, and w are functions of x).

14. Expressed in prime notation, the product rule states that $\big(f(x)g(x)\big)' = \ldots$ *what?*

15. If, for any given x, we can represent the *outputs* of two functions u and v by a rectangle's sides, then we can demonstrate the product rule *geometrically* as follows. Increasing the input by dx yields changes du and dv to the individual outputs, and $d(uv)$ will be represented by an area on the figure. Identify this area, compute it, and divide by dx. You should obtain the product rule.

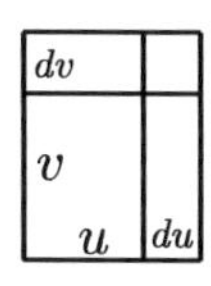

The Quotient Rule

The quotient rule is uglier than the product rule, but it is just as simple to use.

Quotient Rule. If u and v are functions of x, then

$$\frac{d}{dx}\left(\frac{u}{v}\right) = \frac{v\frac{du}{dx} - u\frac{dv}{dx}}{v^2}.$$

Proof. If we increase the input x by an infinitesimal amount dx, then the outputs of the individual functions u and v change by infinitesimal amounts du and dv. Hence, the value of their *quotient* changes from uv to $(u + du)/(v + dv)$. With this in mind, we see that

$$\frac{d\left(\frac{u}{v}\right)}{dx} = \frac{\frac{(u+du)}{(v+dv)} - \frac{u}{v}}{dx} = \frac{\frac{v(u+du) - u(v+dv)}{v(v+dv)}}{dx} = \frac{v(u+du) - u(v+dv)}{v(v+dv)dx} = \frac{vdu - udv}{v^2dx}.$$

Dividing the last expression's top and bottom by dx yields the expression in the box above. ■

The following idiotic jingle will fix the quotient rule in your memory: *Low dee-high minus high dee-low, over the square of what's below.* ("High" and "low" being the top and bottom functions in the quotient, while "dee" indicates "the derivative of".) Recite it when using the quotient rule, and all will be well.

In exercise 9, we employed a devilishly clever geometric argument to prove that $\frac{d}{dx}(\tan x) = \sec^2 x$. With the quotient rule's help, we can prove this fact mechanically, obviating the need for cleverness.

Example 1. Prove that $\frac{d}{dx}(\tan x) = \sec^2 x$.

Solution. Thanks to a well-known trigonometric identity for tangent, we have

$$\frac{d}{dx}(\tan x) = \frac{d}{dx}\left(\frac{\sin x}{\cos x}\right) = \frac{(\cos x)\frac{d}{dx}(\sin x) - (\sin x)\frac{d}{dx}(\cos x)}{\cos^2 x} \quad \text{(by the quotient rule)}$$

$$= \frac{\cos^2 x + \sin^2 x}{\cos^2 x} = \frac{1}{\cos^2 x} = \sec^2 x \quad \text{(by basic trig identities).} \ \blacklozenge$$

The power rule tells us that to take a power function's derivative, we simply reduce its power by 1, and multiply by the old power. (That is, $(x^n)' = nx^{n-1}$.) So far, we've only proved that this holds when the power is a positive integer. With the quotient rule's help, we can show that it holds for *all* integers.

Example 2. Prove that the power rule holds for all integer powers.

Solution. We've already established this for positive integers, and it obviously holds when the power is zero. (Be sure you see why.) To finish the proof, let us suppose $-m$ is a *negative* integer. Then m is a positive integer, so the derivative of x^m is mx^{m-1}. Bearing this in mind, we find that

$$\frac{d}{dx}(x^{-m}) = \frac{d}{dx}\left(\frac{1}{x^m}\right) = \frac{x^m\frac{d}{dx}(1) - 1\frac{d}{dx}(x^m)}{x^{2m}} = \frac{-mx^{m-1}}{x^{2m}} = -mx^{-m-1},$$

which shows that the power rule holds even when the power is a negative integer. ♦

Many problems require us to combine different derivative rules. This is easy enough if you write out the intermediate steps.

Example 3. Find the derivative of $y = \dfrac{x^2 \sin x}{1+x^2}$.

Solution. $\dfrac{dy}{dx} = \dfrac{(1+x^2)\frac{d}{dx}(x^2 \sin x) - (x^2 \sin x)\frac{d}{dx}(x^2)}{(1+x^2)^2}$ (quotient rule)

$= \dfrac{(1+x^2)(2x\sin x + x^2 \cos x) - 2x^3 \sin x}{(1+x^2)^2}$ (product rule). ♦

So much for the quotient rule.

Exercises.

16. Expressed in prime notation, the quotient rule states that $\left(\frac{f(x)}{g(x)}\right)' = \ldots$ *what?*

17. In example 1, we proved that $(\tan x)' = \sec^2 x$. To complete our list of trig functions' derivatives, prove that

a) $(\sec x)' = \sec x \tan x$. b) $(\csc x)' = -\csc x \cot x$. c) $(\cot x)' = -\csc^2 x$.

18. Now that you've found the derivatives of all six trigonometric functions, memorize them. There are patterns (particularly among cofunction pairs) that will make this easier if you notice them. Notice them.

19. Find the derivatives of the following functions (simplifying your answers when possible.)

a) $y = \dfrac{2}{1-x}$ b) $y = \dfrac{3x^3}{\cos x}$ c) $y = \dfrac{x+\pi^2}{\tan x}$ d) $y = \dfrac{x^2}{3x+5}$ e) $y = \dfrac{4x\sin x}{2x + \cos x}$ f) $y = \dfrac{x^2 \sin x}{x \sec x}$

g) $y = \frac{1}{x}(\sec x + \tan x) + \ln 2$ h) $y = x^{-2} - x^{-3} + x^{-4} - x^{-5}$

20. In Example 2, we showed that the power rule holds for all integer exponents. In this exercise, you'll prove that the power rule holds for all *rational* exponents. This will require several steps.

a) First, you'll establish the case where the exponent is a "unit fraction" (i.e. of the form $1/n$ for an integer n). If $y = x^{1/n}$, you can find dy/dx with a diabolical trick: find dx/dy instead and then take its reciprocal! Your problem: Do this.

[*Hint: Begin by expressing x as a function of y. Then compute dx/dy, take its reciprocal, and rewrite it in terms of x. After a little algebra, you should be able to establish that $dy/dx = (1/n)x^{(1/n)-1}$, as claimed.*]

b) Next, having handled the case of a unit fraction exponent, we'll tackle the general rational exponent m/n. The key is to rewrite $y = x^{m/n}$ as the composition of simpler functions whose derivatives we already know. Namely, we can write $y = u^m$, where $u = x^{1/n}$.

Use the trick from exercise 7 (and some algebra) to prove that $dy/dx = (m/n)x^{(m/n)-1}$.

21. Explain how it follows from the previous exercise that $\dfrac{d}{dx}\left(\sqrt{x}\right) = \dfrac{1}{2\sqrt{x}}$.

22. Derivatives of square roots occur often enough to warrant memorizing the formula in the previous exercise. Doing so means that you won't have to rewrite square roots in terms of exponents and apply the power rule each time you need a square root's derivative. After committing this useful formula to memory, find the derivatives of the following functions. (Express your final answers without fractional exponents.)

a) $y = \sqrt{x}\sin x$ b) $y = \sqrt{9x} + \sqrt{x}$ c) $y = \dfrac{\sqrt{x}}{\tan x}$ d) $y = 3\sqrt[3]{x^5}$ e) $y = \csc\left(\sqrt{x}\right)$

23. On the graph of $y = \sin x$, consider two tangent lines: one where $x = 0$, and another where $x = \pi/6$. At which point do these two lines cross?

Derivatives of Exponential Functions

Our quest for the derivative of the exponential function b^x (for any base b) begins algebraically:

$$\frac{d(b^x)}{dx} = \frac{b^{x+dx} - b^x}{dx} = \frac{b^x b^{dx} - b^x}{dx} = \left(\frac{b^{dx} - 1}{dx}\right) b^x.$$

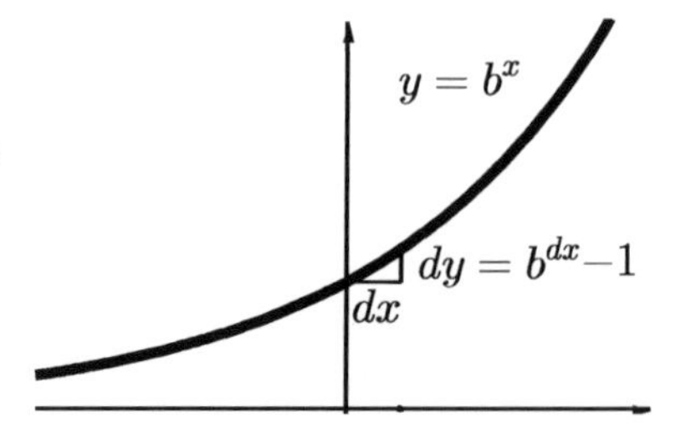

The expression in parentheses is, as the figure shows, the slope of b^x at (0,1). Consequently, we can rewrite the preceding equation as

$$\frac{d(b^x)}{dx} = \begin{pmatrix} \text{The slope of } b^x \\ \text{at point } (0,1) \end{pmatrix} b^x.^*$$

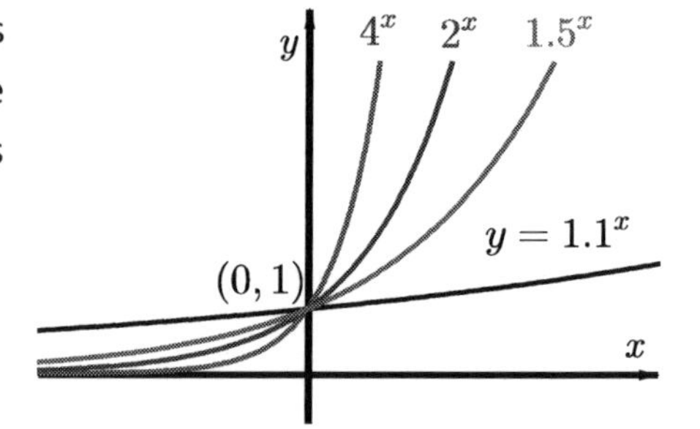

As the next figure demonstrates, increasing the base b increases the slope. Moreover, *some* base (between 1.1 and 4) will make the slope at (0,1) equal to 1 exactly. We call this special base $\boldsymbol{e}$. This endows the function e^x with a truly remarkable property:

$$\frac{d}{dx}(e^x) = (1)e^x = e^x.$$

That is, the function e^x is its own derivative.

No doubt you are wondering if this is the same e you've met before as the natural logarithm's base. It is. In precalculus, e is enigmatic. In calculus, e arises *naturally*, since, by its very definition, it is the base of the one exponential function equal to its own derivative.† Perhaps e should not be introduced in precalculus classes at all; grasping e's significance before knowing what a derivative is may be as hopeless as understanding π's significance before knowing what a circle is.

Now that we know the derivative of e^x, we can find the derivative of b^x for *any* base b. We begin by using logarithmic properties to convert an arbitrary-based exponential function to one with base e.

$$b^x = \left(e^{\ln b}\right)^x = e^{(\ln b)x}.$$

Next, we'll use a trick that you've already used in several exercises (see exercises 7, 20b, and 22e). Namely, we shall rewrite b^x as a composition of simpler functions whose derivatives we already know.

$$b^x = e^u, \text{ where } u = (\ln b)x.$$

This allows us to compute the derivative we seek:

$$\frac{d(b^x)}{dx} = \frac{d(b^x)}{du}\frac{du}{dx} = (e^u)(\ln b) = (\ln b)e^{(\ln b)x} = (\ln b)\left(e^{(\ln b)}\right)^x = (\mathbf{ln}\, \boldsymbol{b})\boldsymbol{b}^{\boldsymbol{x}}.$$

---

* For example, at (0,1), the graph of $y = 2^x$ has a slope of approximately 0.7, so it follows that $\frac{d}{dx}(2^x) \approx 0.7(2^x)$.

† Stated with a bit more care, we can define e is the base of the only function **of the form $\boldsymbol{b^x}$** that is equal to its own derivative. It is reasonable to ask if there any *other* functions (of another form) that equal their own derivatives. It turns out that e^x is essentially unique in this regard; the only other such functions that have this property are constant multiples of e^x.

We have thus established the following results:

Derivatives of Exponential Functions.

$$\frac{d}{dx}(e^x) = e^x.$$

More generally, for any positive base b,

$$\frac{d}{dx}(b^x) = (\ln b)b^x.$$

We saw earlier (in exercise 4) that we use radians precisely because they simplify sine's derivative; if calculus didn't exist, neither would radians. Similarly, we use e as a base for exponential functions and logarithms because it simplifies their derivatives. If calculus did not exist, no one would bother with e. Calculus forces strange things up from the depths.

When exponential functions are present, their inverses – logarithms – are never far away. Once we know a function's derivative, we can find its inverse's derivative (by using the trick from exercise 20a). I'll use that trick here to find the natural logarithm's derivative. (And you'll use the trick in the exercises to find the derivatives of the base-10 logarithm and the inverse trigonometric functions.)

Problem. Find the derivative of $y = \ln x$.

Solution. Since $y = \ln x$ implies $x = e^y$, we have $\frac{dx}{dy} = e^y$. Thus, $\frac{dy}{dx} = \frac{1}{e^y} = \frac{1}{e^{\ln x}} = \frac{1}{x}$. ♦

The natural logarithm's derivative is important. Commit it to memory.

$$\frac{d}{dx}(\ln x) = \frac{1}{x}.$$

Exercises.

24. Find the following functions' derivatives.

a) $y = e^x \sin x$ b) $y = \frac{\ln x}{3\tan x}$ c) $y = -2^x \cos x + 3x^2 \ln x$ d) $f(x) = e^x\sqrt{x} + e^2$

e) $g(x) = 2^x 3^x 4^x 5^{x+1}$ [*Hint: Some preliminary algebra will help.*] f) $y = \csc x \sin x - \frac{\ln e^x}{x}$

g) $w = \sqrt[3]{t^5}\sec t + \frac{1}{t}$ h) $V = \frac{1}{y} + \ln y$ i) $k(x) = \frac{2^x \cos^2 x + 2^x \sin^2 x}{\sqrt{x}}$ j) $y = x^e/e^x$

25. Does the graph of $y = 2^x - x$ have a horizontal tangent at any point? If so, find that point's x-coordinate.

26. Find the derivative of $y = \log_{10} x$ by adapting the trick we used to establish the natural logarithm's derivative.

27. The same trick can be used to find the derivative of $y = \sin^{-1} x$, but a trigonometric twist complicates the end.

a) If you solved the previous problem, you'll find it easy to show that $dy/dx = 1/\cos(\sin^{-1} x)$. Do so.

b) We can simplify the ghastly expression $\cos(\sin^{-1} x)$ with a trigonometric identity. To derive it, first explain why $\cos(\sin^{-1} x) = \pm\sqrt{1 - \sin^2(\sin^{-1} x)}$. Then simplify this and explain why the $\pm$ must in fact be a plus. [*Hint for the* ± business: *Think about the range of inverse sine, and what cosine does to values in that range.*]

c) Conclude that $\frac{d}{dx}(\sin^{-1} x) = \frac{1}{\sqrt{1-x^2}}$.

28. By shadowing the argument we used in the previous exercise, show that $\frac{d}{dx}(\cos^{-1} x) = \frac{-1}{\sqrt{1-x^2}}$.

29. Finally, show that $\frac{d}{dx}(\tan^{-1} x) = \frac{1}{1+x^2}$. [*Hint: Use the alternate Pythagorean identity,* $\sec^2 x = 1 + \tan^2 x$.]

30. In this problem, you'll learn how to approximate the numerical value of e.

a) Explain how e's definition implies that $(e^{dx} - 1)/dx = 1$, which, in turn, implies that $e = (1 + dx)^{1/dx}$. [*Hint: Express the slope of the graph of* $y = e^x$ *at* $(0,1)$ *two different ways. Equate the results.*]

b) The formula for e in part (a) is approximately true if we replace the infinitesimal dx by a small *real* value Δx. The smaller the value of Δx, the better the approximation. Using a calculator, substitute small values of Δx into the approximation $e \approx (1 + \Delta x)^{1/\Delta x}$, and convince yourself that $e \approx 2.71828$.

c) One way to make Δx small is to let $\Delta x = 1/n$, where n is a large whole number. If we make this substitution, then we can reformulate the approximation in the previous part as follows: $e \approx (1 + 1/n)^n$, where n is a large whole number; the larger n is, the better the approximation. Using a calculator, substitute some large values of n into this approximation to confirm what you discovered in part (b).

31. In #27, some clever algebraic shenanigans led you to *inverse sine*'s derivative. Since inverse sine, however, is an essentially geometric function, geometrically-minded souls will find the following geometric derivation more illuminating.

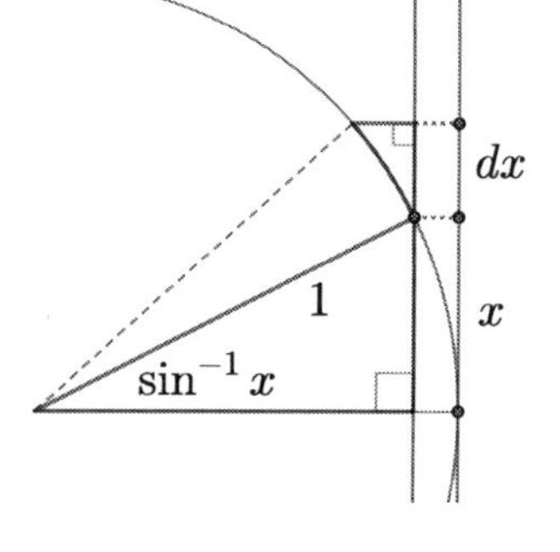

a) Defining x as in the figure, explain why the angle marked $\sin^{-1} x$ has been marked appropriately.

b) Increasing x by an infinitesimal amount dx induces an infinitesimal change in $\sin^{-1} x$. Locate $d(\sin^{-1} x)$ on the figure.

c) One leg of the figure's infinitesimal triangle is in fact an arc. Find its length.

d) Explain why the two right triangles emphasized in the figure are similar.

e) Use this similarity to show that $d(\sin^{-1} x)/dx = 1/\sqrt{1 - x^2}$.

32. In light of the previous problem, it should come as no surprise that the other inverse trigonometric functions' derivatives can be justified geometrically. In this problem, you'll carry this out for inverse tangent.

a) Explain why the angle marked $\tan^{-1} x$ has been appropriately marked.

b) Locate $d(\tan^{-1} x)$ on the figure.

c) Use the Pythagorean theorem to find the large right triangle's hypotenuse.

d) One leg of the figure's infinitesimal triangle is in fact an arc. Find its length.

e) Explain why the two right triangles emphasized in the figure are similar.

f) Use this similarity to show that

$$\frac{d(\tan^{-1} x)}{dx} = \frac{1}{1 + x^2}.$$

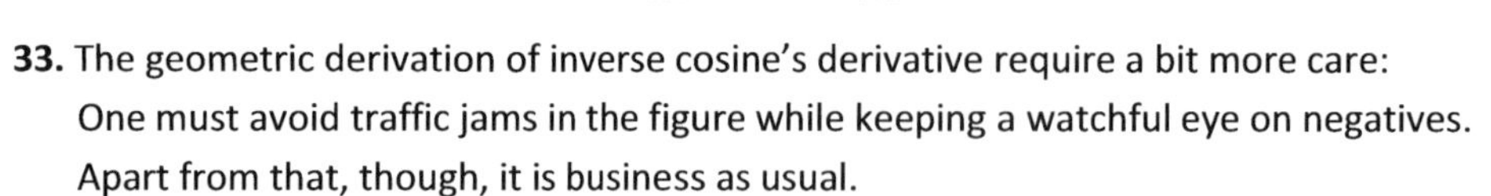

33. The geometric derivation of inverse cosine's derivative require a bit more care: One must avoid traffic jams in the figure while keeping a watchful eye on negatives. Apart from that, though, it is business as usual.

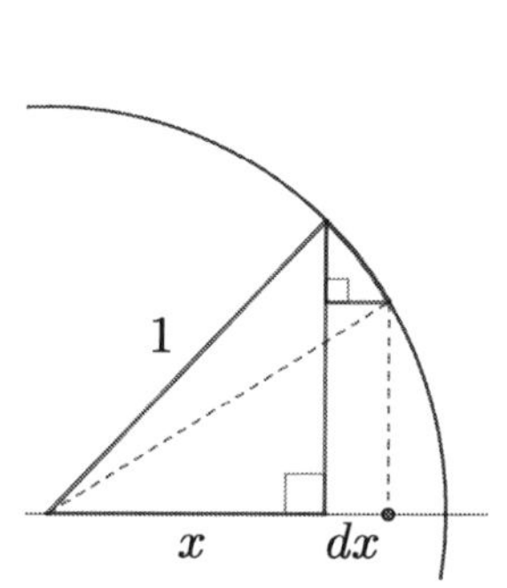

a) Defining x as in the figure, locate $\cos^{-1} x$.

b) Observe that increasing x by an infinitesimal amount dx induces an infinitesimal *decrease* in $\cos^{-1} x$, so $d(\cos^{-1} x)$ is negative. Locate the quantity $-d(\cos^{-1} x)$, which is *positive*, on the figure.

c) Complete the argument.

The Chain Rule

"He hath hedged me about, that I cannot get out: he hath made my chain heavy."
- Lamentations 3:7

You have already met the chain rule, for I smuggled it – incognito – into some earlier exercises (7, 20b). The rule is really nothing but the trivial observation that we can factor dy/dx as follows:

$$\frac{dy}{dx} = \frac{dy}{du} \cdot \frac{du}{dx}.$$

To find derivatives of *composite functions* (i.e. functions of functions, such as $y = \sin(\ln x)$), we just call the composite's "inner function" u, then grind out the two derivatives on the right-hand side above. Note that the first factor, dy/du, will be a function of u, so after computing it, we must rewrite it in terms of x, which is a simple matter of substituting the inner function itself for u wherever it appears.

Example 1. Find the derivative of $y = \sin(\ln x)$.

Solution. If we rewrite this function as $y = \sin u$, where $u = \ln x$, the chain rule gives

$$\frac{dy}{dx} = \frac{dy}{du} \cdot \frac{du}{dx} = \cos u \left(\frac{1}{x}\right) = \frac{\cos(\ln x)}{x}. \quad \blacklozenge$$

Note that in our final expression, we've eliminated all traces of u.

Example 2. Find the derivative of $y = (3x^2 - 1)^{10}$.

Solution. If we rewrite this as $y = u^{10}$, where $u = 3x^2 - 1$, the chain rule yields

$$\frac{dy}{dx} = \frac{dy}{du} \cdot \frac{du}{dx} = 10u^9(6x) = 60x(3x^2 - 1)^9. \quad \blacklozenge$$

We can speed up the chain rule procedure (and make it independent of Leibniz notation) by restating it in words. To do so, we first note that dy/du and du/dx are the derivatives of the composite's outer and inner functions, respectively. We can now state the "fast version" of the chain rule: **Take the outer function's derivative, *evaluate it at the inner function*, then multiply by the inner function's derivative**. (The italicized step corresponds to eliminating u from the final expression.)

Here's an example of the fast version in action.

Example 3. Find the derivative of $y = \tan(2x^3 + 3x^2 + x + 1)$.

Solution. By the chain rule, $\frac{dy}{dx} = \underbrace{\left(\sec^2(2x^3 + 3x^2 + x + 1)\right)}_{\substack{\text{Derivative of the outer} \\ \text{evaluated at the inner.}}} \underbrace{(6x^2 + 6x + 1)}_{\substack{\text{Derivative} \\ \text{of the inner}}}$ ♦

This fast version of the chain rule is especially useful when the chain rule must be combined with the product rule, the quotient rule, or even a second instance of the chain rule.

Example 4. Find the derivative of $y = \sqrt{3x^2 \sin x}$.

Solution.
$$\frac{dy}{dx} = \frac{1}{2\sqrt{3x^2 \sin x}} \cdot \frac{d}{dx}(3x^2 \sin x) \qquad \text{(chain rule)}$$
$$= \frac{1}{2\sqrt{3x^2 \sin x}}[6x \sin x + 3x^2 \cos x] \qquad \text{(product rule)} \quad \blacklozenge$$

For our grand finale, a fourfold composition of functions. It looks more complicated, but isn't really; we just follow the chain of derivatives, link by link, from the outermost to the innermost function.

Example 5. Find the derivative of $y = \sin(\tan(\ln(2x+2)))$.

Solution. $\frac{dy}{dx} = \cos(\tan(\ln(2x+2)))\frac{d}{dx}(\tan(\ln(2x+2)))$ (chain rule)

$= \cos(\tan(\ln(2x+2)))\,(\sec^2(\ln(2x+2))\frac{d}{dx}(\ln(2x+2))$ (chain rule)

$= \cos(\tan(\ln(2x+2)))\,(\sec^2(\ln(2x+2))\left(\frac{1}{2x+2}\right)\frac{d}{dx}(2x+2)$ (chain rule)

$= \cos(\tan(\ln(2x+2)))\,(\sec^2(\ln(2x+2))\left(\frac{1}{x+1}\right)$. ♦

Mastering the chain rule requires practice. Behold your opportunity to obtain it:

Exercises.

34. Express the chain rule in prime notation.

35. Find the derivatives of the following functions.

a) $y = e^{5x}$ b) $y = \cos(3x^2) + 1$ c) $y = \ln(\ln x)$ d) $y = (2x^3 - x)^8$ e) $y = 2^{\tan x}$

f) $y = \frac{1}{(3x-4)^2}$ g) $y = x\sqrt{144 - x^2}$ h) $y = e^{-x^2}$ i) $y = \sqrt[3]{x + x^3}$ j) $y = \cos^2 x$

k) $y = (1 - 4x^3)^{-2}$ l) $y = \cos(\csc(1/x))$ m) $y = e^{\sin^2 x}$ n) $y = \ln(\sin(\ln x))$ o) $y = \frac{e^{\sin x}}{\sqrt{\sec x}}$

p) $y = \ln\left(\frac{x^2+4}{2x+3}\right)$ q) $y = 10^{-x/(1+x^2)} + 2^{\pi}$ r) $y = 5x^2 e^{6x^2+e}$ s) $y = \ln\left(\sqrt[3]{6x^2+3x}\right)$ t) $y = e^{\pi^2}$

36. The 27th-degree polynomial $y = (2x^3 - x^2 + x + 1)^9$ crosses the y-axis at some point. Consider the tangent to the polynomial's graph at this point. Find the point at which this tangent line crosses the x-axis.

37. Does the graph of $y = \ln(\sin(\ln x))$ have a horizontal tangent line at any of its points? If so, at how many points? Find the exact coordinates of one such point.

38. Let $\sin_{<n>}(x)$ denote n nested sine functions, so that, for example, $\sin_{<3>}(x) = \sin(\sin(\sin x))$.

a) Where does the graph of $y = \sin_{<2>}(x)$ cross the x-axis?

b) What is the range of $y = \sin_{<2>}(x)$? State it exactly, then use a calculator to approximate its endpoints.

c) What is the slope of $y = \sin_{<2>}(x)$ at the origin? At other points where the graph crosses the x-axis?

d) Repeat the first three parts of this problem, but with $y = \sin_{<3>}(x)$.

e) Graph $y = \sin_{<2>}(x)$ and $y = \sin_{<3>}(x)$ as best you can. Then try to guess what happens to the graphs of the functions $y = \sin_{<n>}(x)$ for larger and larger values of n.

Chapter 3

Applications – Contrived and Genuine

Concavity and the Second Derivative

If a function's value is increasing throughout an interval, its derivative must be positive throughout that interval, since a function's value can increase only where its slope is positive. Similarly, if a function's value decreases throughout an interval, its derivative must be negative over that same interval. In short, the sign of a function's derivative tells us where the function's graph is going up or going down.

Interestingly, the sign of a function's *second* derivative also tells us something about the function's graph: It can be shown that a function's second derivative is positive over intervals where the function's graph is "concave up" (like a cup*), and negative over intervals where the function's graph is "concave down" (like a frown).

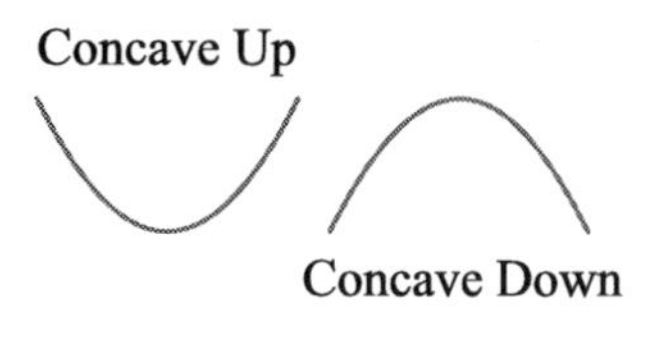

Why is this so? The explanation is a bit subtle. I will convey its basic flavor in the following paragraph.

The key idea is that a function's *second* derivative gives the rate at which its first derivative changes. Or, to restate this geometrically, **the second derivative gives the rate at which a function's *slope* changes.** Thus, a function whose *slope* increases throughout an interval (such as f and g at right) must have a positive second derivative throughout that interval.† Similarly, a function whose slope decreases throughout an interval (such as h and k) must have a negative second derivative there. Observe that f and g, the two functions in the figure with positive second derivatives (i.e. with increasing slopes) are indeed concave up. Moreover, h and k, the two functions with negative second derivatives (decreasing slopes) are concave down. Should you care to indulge in a little meditation upon this idea, you'll soon be convinced that this link between concavity and the second derivative holds for all functions.

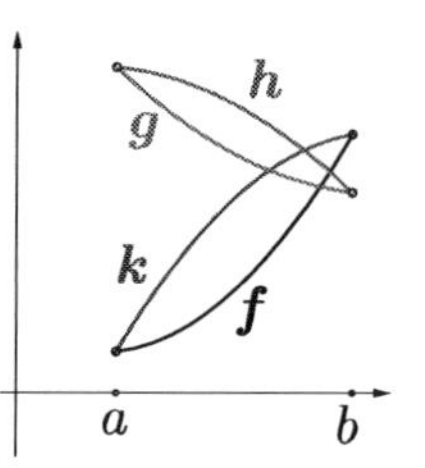

For future reference, here are the essential results summarizing the geometric relationships that hold among a function's graph and its first *two* derivatives:

The graph of f is increasing [or decreasing] on an interval.	$\Leftrightarrow$	**$f'(x)$ is positive** [negative] for all x in the interval.
The graph of f is concave up [or concave down] on an interval.	$\Leftrightarrow$	**$f''(x)$ is positive** [negative] for all x in the interval.

Finally, a bit of terminology: Any point at which a graph switches from concave down to concave up (or vice-versa) is called an *inflection point* of the graph. It is visually obvious that the graph of $f(x) = x^3$ is concave *down* over the negative reals, and concave *up* over the positive reals, so it has an inflection point at the origin. Even if we didn't already know this function's graph, we still could have found this inflection point analytically by noting that $f''(x) = 6x$, which is negative when $x < 0$, and positive when $x > 0$.

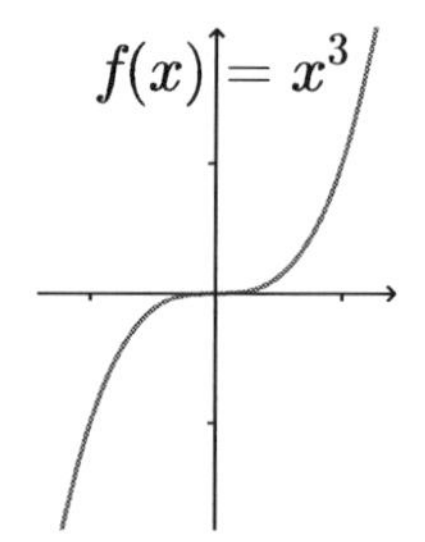

* Or *part of* a cup. A comma, for example, is concave up. If a comma were the graph of a function, its second derivative would be positive everywhere it is defined.

† Near a, the slope of f is positive but close to zero; as we move from left to right towards b, its slope climbs above 1. Similarly, the slope of g starts at a negative value, but increases to a negative value that is closer to zero.

Exercises.

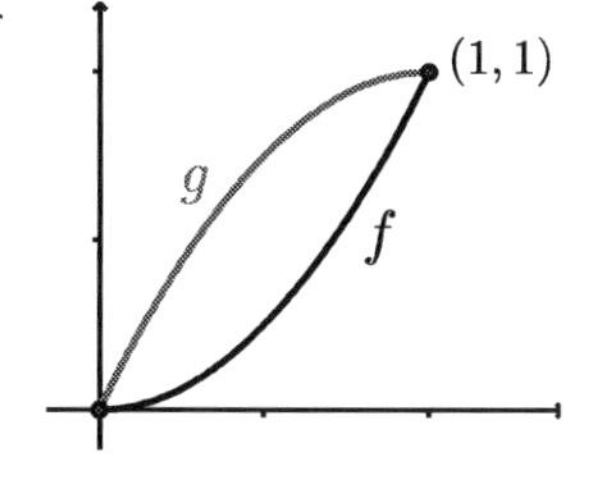

1. The figure at right shows two functions defined on the interval $(0,1)$. State whether each of the following statements is true or false.

a) $f(x) > 0$ for all x in $(0,1)$. b) $f'(x) > 0$ for all x in $(0,1)$.
c) $f''(x) > 0$ for all x in $(0,1)$. d) $g(x) > 0$ for all x in $(0,1)$.
e) $g'(x) > 0$ for all x in $(0,1)$. f) $g''(x) > 0$ for all x in $(0,1)$.

g) f has an inflection point. h) g has an inflection point.

2. By thinking about graphs of the basic functions you know, find examples that are...

a) increasing and concave up on $(0,1)$. b) increasing and concave down on $(0,1)$.
c) decreasing and concave up on $(0,1)$. d) decreasing and concave down on $(0,1)$.

3. Regarding your answers to the previous exercise, verify that over the interval $(0,1)$, the increasing functions do indeed have positive derivatives, the decreasing functions have negative derivatives, those with concave up graphs have positive second derivatives, and those with concave down graphs have negative second derivatives.

4. Every "hill" or "valley" in a function's graph has a horizontal tangent at its extremity, but the converse is *not* true; a graph can have a horizontal tangent at a point that is neither hill nor valley.

a) Verify that this last possibility occurs on the graph of $y = 3$.
b) More interestingly, verify that it also occurs at one point on the graph of $y = x^3$.
c) For a function's graph to have a hill or valley at a point, the function's derivative must not only equal zero at that point, but it must also *change sign* there (from positive before to negative after – or vice-versa). Explain intuitively why this is so. Use your explanation to deepen your understanding of the example in part (b).

5. Every inflection point in a function's graph occurs where the second derivative is zero, but the converse is *not* true; a graph can have a point at which the second derivative equals zero, and yet is not an inflection point.

a) Verify that this last possibility occurs on the graph of $y = x^4$ at the origin.
b) For a function's graph to have an inflection point, the function's second derivative must not only equal zero at that point, but it must also... *what*?

6. Leibniz's notation for the second derivative of y with respect to x is awkward: $\boldsymbol{d^2y/dx^2}$. This is one of those places where prime notation has the advantage, which is why I used it in the preceding section. Still, you should know the Leibniz notation for the second derivative, since you will meet it elsewhere if you continue to study mathematics and physics. The idea behind the notation is that since y's derivative is dy/dx, its second derivative (i.e. the derivative *of its derivative* with respect to x) is, of course,

$$\frac{d}{dx}\left(\frac{dy}{dx}\right),$$

which is compressed (for better or for worse) to

$$\boldsymbol{\frac{d^2y}{dx^2}}.$$

(Thus, for example, if $y = 3x^3$, it follows that $\frac{dy}{dx} = 9x^2$ and $\frac{d^2y}{dx^2} = 18x$.) Normally, a major advantage of Leibniz notation over its competitors is that it keeps the derivative's true nature – a ratio of infinitesimals – before us at all times. This is not true, however, for the Leibniz notation for *second* derivatives: d^2y/dx^2 does *not* represent the ratio of its alleged numerator and denominator, neither of which has individual meaning. (In particular, the "denominator" is certainly *not* the product dx and dx!) For this reason, I'll rarely use Leibniz notation for second (or higher-order) derivatives. For higher-order derivatives, the notation generalizes in the obvious way: the *third* derivative of y with respect to x is denoted d^3y/dx^3, and so forth.

a) If $y = \sin x$, what is $\frac{d^2y}{dx^2}$? b) If $y = \tan x$, what is $\frac{d^2y}{dx^2}$? c) If $y = e^x$, what is $\frac{d^{913}y}{dx^{913}}$?

d) If $z = \pi^w$, what is $\frac{d^2z}{dw^2}$? e) If $z = \sin t$, what is $\frac{d^{913}z}{dt^{913}}$? f) If $y = \ln x$, what is $\left(\frac{d^3y}{dx^3}\right)x^3$?

g) True or false: If $y = p(x)$ is a polynomial function, there is some whole number n such that $\frac{d^ny}{dx^n} = 0$.

7. You've probably learned that a polynomial's *degree* tells us the maximum number of zeros it can have. (For example, a quadratic can have at most two zeros.) Use this fact to explain why the maximum number of "turns" in a polynomial's graph (*i.e.* where it turns from increasing to decreasing or vice-versa) is *its degree minus one*.

8. A central concept of **differential geometry** (a subject you may study someday) is *curvature*, a measure of "how curved" a geometric object is at a given point. It turns out that two functions' graphs have the same curvature at the same point precisely when the functions agree on their first two derivatives there. [In symbols, f and g have the same curvature at a if and only if $f'(a) = g'(a)$, and $f''(a) = g''(a)$.]

If two curves passing through the same point have the same curvature there, then in the vicinity of that point, the curves follow one another's movements with remarkable precision: Not only do they pass through the same point, but their values – and even their slopes – change at exactly the same rate as they pass through it. They are like two swallows in an aerial mating ritual. Consequently, the simpler of the two functions can be used to approximate the value of the more complicated function for nearby points.

a) Verify that $f(x) = \sin x$ and $g(x) = x$ pass through $(0,0)$ and have the same curvature there. Deduce that for values of x close to 0, $f(x) \approx g(x)$, and use this to approximate the value of $\sin(0.05)$ by hand. Check your approximation with a calculator.

b) Verify that $f(x) = e^x$ and $g(x) = 1 + x + (x^2/2)$ pass through $(0,1)$ with the same curvature. Use this fact to approximate the value of $e^{0.07}$ by hand. Check your approximation with a calculator.

c) Show that $f(x) = e^x$ and $h(x) = 1 + x + (x^2/2) + (x^3/6) + (x^4/24)$ agree at 0 up to the *fourth* derivative. [*i.e.* Show that $f(0) = h(0)$, $f'(0) = h'(0)$, $f''(0) = f''(0)$, $f'''(0) = h'''(0)$, and $f''''(0) = h''''(0)$.] Then use h to approximate $e^{0.07}$ by hand. With a calculator, confirm that this approximation is accurate in the first seven decimal places. The synchronized dance of this particular pair of swallows is astonishingly accurate.

d) When you used your calculator to compare your approximations to the "true" value of $e^{0.07}$ in the two previous parts, how did your calculator know the "true" value? The short answer is that *it didn't*. It too was using a polynomial to produce its approximation – though its polynomial may have agreed at zero with e^x not just on their first 2 or 4 derivatives, but perhaps on their first 20. In light of the two previous parts, can you guess the coefficient of this polynomial's x^5 term?

9. Cubic functions have the form $y = ax^3 + bx^2 + cx + d$. Their graphs come in two basic varieties: those with two bumps, and those with none.

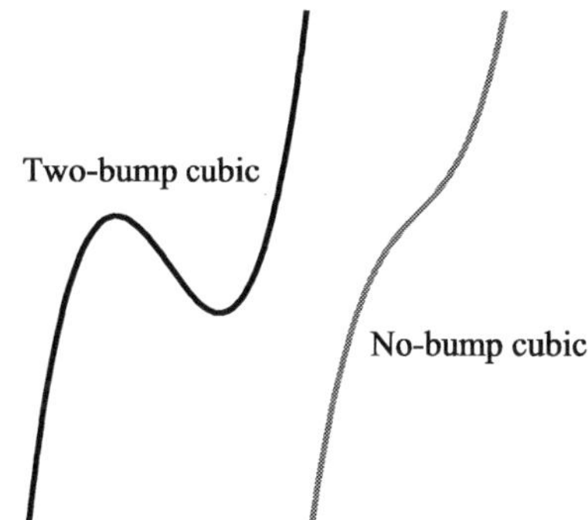

a) Why are there no "one-bump" cubics?

b) Find a "bump discriminant": a purely algebraic test that distinguishes the two-bumpers from the no-bumpers.

c) Every cubic has an inflection point. Find a formula for its x-coordinate. (The formula will be in terms of the cubic's coefficients, of course.)

d) Find the inflection point of $f(x) = x^3 + 9x^2 + 25x + 21$. If we shift the graph horizontally so that its inflection point lies on the y-axis, the shifted graph will have a new equation. Find it, and note that this equation lacks an x^2 term. This isn't an accident.

e) Prove that if you shift any cubic so that its inflection point lies on the y-axis, the equation of the shifted cubic will lack an x^2 term. Such a cubic is called a *depressed cubic*. (Pining for its lost x^2 term, no doubt.)

f) Depressed cubics play a role in a particularly fascinating episode in the history of mathematics. We begin in the early 16th century, when Scipione del Ferro figured out how to solve depressed cubic equations. In particular, del Ferro discovered that one solution to the general depressed cubic $x^3 + cx + d = 0$ is

$$x = \sqrt[3]{-\frac{d}{2} + \sqrt{\frac{d^2}{4} + \frac{c^3}{27}}} + \sqrt[3]{-\frac{d}{2} - \sqrt{\frac{d^2}{4} + \frac{c^3}{27}}}.$$

I omit the details of del Ferro's derivation, as they would take us too far afield.* The fascinating episode I have in mind occurred in 1572, when Rafael Bombelli studied the depressed cubic equation $x^3 - 15x - 4 = 0$. To begin, he noted that it has three solutions: 4 and $-2 \pm \sqrt{3}$. (Please verify that these are indeed solutions.) Now, del Ferro's formula produces a solution to *any* depressed cubic, so applying it to Bombelli's depressed cubic must yield one of the three solutions noted above, for a cubic can have no more than three solutions. **Your problem:** Apply del Ferro's formula to Bombelli's cubic and see what emerges.

g) Remind yourself *why* negative numbers are supposedly incapable of having square roots.

h) When the *quadratic* formula yields a "solution" involving the square root of a negative, we can read it as an error message indicating that the quadratic actually has *no solution*. However, we can't do this with the result you found in (f), for we *know* that Bombelli's cubic has solutions! Struggling at this impasse, Bombelli had what he called "a wild thought": Suppose, just suppose, we took square roots of negatives seriously, not as error messages but as genuine numbers, despite (g). **Your problem:** Indulge Bombelli and show that cubing $2 + \sqrt{-1}$ yields $2 + 11\sqrt{-1}$. Then show that cubing $2 - \sqrt{-1}$ yields $2 - 11\sqrt{-1}$, whatever that might mean.

i) Still working in Bombelli's fantasy, explain how your calculations in the previous part imply that

$$2 + \sqrt{-1} = \sqrt[3]{2 + 11\sqrt{-1}} \text{ and } 2 - \sqrt{-1} = \sqrt[3]{2 - 11\sqrt{-1}},$$

Which in turn implies that the solution to Bombelli's formula that you found in part (f) with del Ferro's solution is in fact **4** in disguise! Bombelli's wild thought is vindicated! This was how complex numbers arrived in mathematics: They were born not of paradoxical speculation, but of the need to *resolve* a paradox.

* I note in passing, however, that we can actually use del Ferro's formula to find a zero of *any* cubic polynomial: First, we "depress" the cubic by shifting its graph as in part (e), then we find an x-intercept of the shifted graph via del Ferro's formula; finally, we translate this intercept back to its location on the original graph. By carrying this out on the general cubic, we could even derive a "cubic formula" akin to the "quadratic formula" you know so well. If you try to carry this out yourself, be sure to use jumbo-sized paper. There is a reason why you were taught how to solve quadratics, but not cubics.

Optimization

"…Carthage, the new town. They bought the land,
Called Drumskin from the bargain made, a tract
They could enclose with one bull's hide."
- Virgil, *The Aeneid*, Book I (tr. Robert Fitzgerald)

Those who have read Virgil will remember Dido's doomed love for Aeneas. Those who have not should. Virgil's lines above are his sole reference to a still older story: Exiled from her native city, Dido and her followers made their way to the North African coast, where she convinced the local king to sell her an ostensibly miniscule bit of territory – just as much land as she could enclose in a bull's hide. Dido revealed her foxy genius *after* sealing the deal, when she sliced the hide into thin strips, sewed these end to end, and arranged the resulting "rope" into a circle large enough to surround the hill that thus became the site of her new city, Carthage. But why a *circle*? Why not, say, a square? The answer is simple: Among all shapes with the same fixed perimeter (in Dido's case, the fixed length of her rope), a circle encompasses the greatest area. Dido's choice of a circle thus maximized her territory.

Differential calculus is a natural tool for **optimization** – maximizing or minimizing a variable quantity (profit, pollution, or whatnot). A function's maximum and minimum values generally correspond to the tops and bottoms of hills and valleys in its graph; these have horizontal tangents, which, of course, occur where the function's derivative is zero. This simple idea can help us solve optimization problems.

Example 1. Suppose we must design a water tank in the shape of a topless, square-based box, and that the tank must hold exactly 1000 cubic inches of water. If we wish to minimize the amount of material used for the box, what should its dimensions be?

Solution. Our goal is to minimize *surface area*, so we'll begin by writing down an algebraic expression for surface area in terms of the labels on the figure:

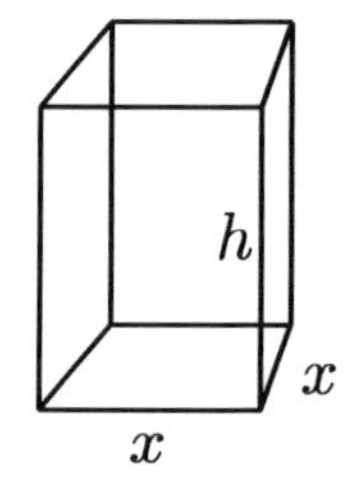

$$A = x^2 + 4hx.$$

This seems to be a function of *two* variables, but we can eliminate one. Equating two expressions for the box's volume ($x^2h = 1000$), we obtain $h = 1000/x^2$. Substituting this into our area formula yields

$$A(x) = x^2 + (4000/x)\,.$$

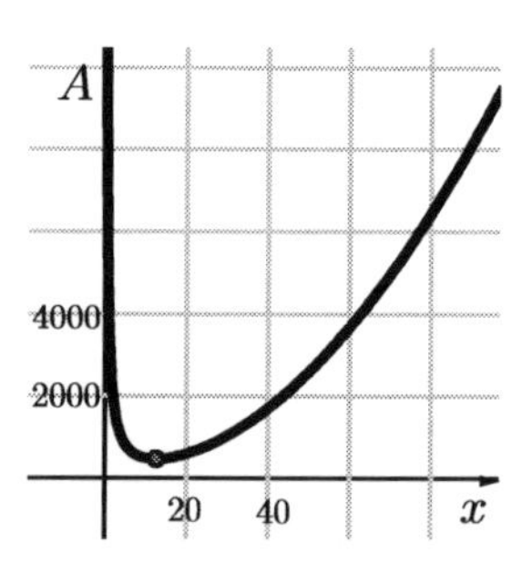

We haven't used any calculus so far. In fact, if we are content with an *approximate* answer (if, say, this were a real engineering problem), then calculus is unnecessary even to *finish* the problem. We would just graph the area function on a computer, and observe that A is minimized when $x \approx 12.6$. (For greater accuracy, we just zoom in closer.) For this x-value, $h \approx 1000/(12.6)^2 = 6.3$, so to minimize the box's surface area, we should give it a 12.6" × 12.6" base and make it 6.3" high.

For an *exact* solution, we need calculus. We note that A is minimized when $A'(x) = 0$. Solving this equation yields $x = 10\sqrt[3]{2}$, as you should verify. Hence, the sides of the optimal box's base are exactly $10\sqrt[3]{2}$ inches, and the box's height is $h = \frac{1000}{\left(10\sqrt[3]{2}\right)^2} = 5\sqrt[3]{2}$ inches. ♦

The preceding example raises two questions. First, in an applied problem (as this one purports to be), decimal approximations are both sufficient and necessary, so why bother with calculus? Who needs an exact answer? Why not just graph the damned thing and be done with it?

Second, suppose your cruel calculus teacher forbids calculators on exams. Without seeing A's graph, how could you know that the horizontal tangent at $x = 10\sqrt[3]{2}$ occurs at a valley, where area is minimized? Maybe it occurs at a hilltop, where area is maximized! (Or at neither hill nor valley, as in exercise 4.) Without a graph, how can you distinguish minima from maxima?

Let us answer these questions in order.

First, with help from a graphing program, a good *pre*calculus student can indeed solve just about any "applied optimization" problem in freshman calculus.* This, however, is only because freshman calculus courses confine themselves, of necessity, to baby functions: functions of a *single variable*. Genuine applied problems involve functions of *multiple* variables. (This is why textbook "applied" problems are so patently phony.) The graphs of baby functions are curves in the plane; we can, of course, plot these on a computer screen and literally *see* their hills and valleys. The graphs of functions of *two* variables are surfaces in three-dimensional space; with more sophisticated graphing programs, we can see their extrema, too. But the graphs of functions of *three or more* variables lie in spaces of *four or more dimensions*, which means they are completely inaccessible to graphing programs – now and forever. Calculus laughs at these limitations; unlike graphing programs, it works perfectly well in spaces of four, five, even ten thousand dimensions. To understand how optimization works in **multivariable calculus**, however, you must first understand how it works in single-variable calculus. And this, O reader, is why you'll use calculus to solve "applied" optimization problems in the present course. One does not learn calculus to solve dopey applied problems. One solves dopey applied problems to learn calculus.

Now for the second question: Without a graph, how can we tell if a horizontal tangent corresponds to a hill, a valley, or neither? Sometimes we can sketch a graph by hand, in which case the problem vanishes.† In other cases, context makes the answer obvious. And at still other times, we can distinguish hills from valleys by thinking about the function's *second derivative*. If at some point c, a function has a horizontal tangent *and is concave up* (i.e. has a positive second derivative), then it should be obvious to you that the function's graph must have a valley at c. (Take a moment or two to convince yourself that this *is* obvious.) For instance, in our preceding example, a little calculation shows that $A''(x) = 2 + (8000/x^3)$, from which it follows that $A''(10\sqrt[3]{2}) > 0$; hence, the area function is concave up at $10\sqrt[3]{2}$, which means that the horizontal tangent there must correspond to a valley. The function is therefore *minimized*. This trick for distinguishing hills from valleys is often called **the second derivative test**. Naturally, it cuts both ways: If the tangent is horizontal where the graph is concave *down*, then the graph will clearly have a hilltop (maximum) there.

---

* Do bear in mind, though, that most computer graphing programs can be flummoxed by even simple functions. Any algebra student can graph $y = (40^x)(50^x) - 2000^x$ by noting that it simplifies to $y = 0$. Most graphing programs, however, cannot. Try graphing this function on a few programs and behold the spectacle of "overflow error," a particular form of computer obtuseness. (In 1996, overflow error led to the destruction of the European Space Agency's $370 million *Cluster* spacecraft.)

† This would work in the preceding example, where $A(x)$ was the sum of x^2, a parabola, and $4000/x$, a vertically-stretched reciprocal function. Graph them, "add" them (i.e. add their corresponding y-coordinates), and you'll have the graph of $A(x)$.

Example 2. Of all points on the graph of $y = \sqrt{x}$, find the one closest to $(3,0)$.

Solution. Let $(x, \sqrt{x})$ be an arbitrary point on the graph. By the familiar distance formula, its distance s from $(3,0)$ is given by

$$s = \sqrt{(x-3)^2 + \left(\sqrt{x} - 0\right)^2} = \sqrt{(x-3)^2 + x}\,.$$

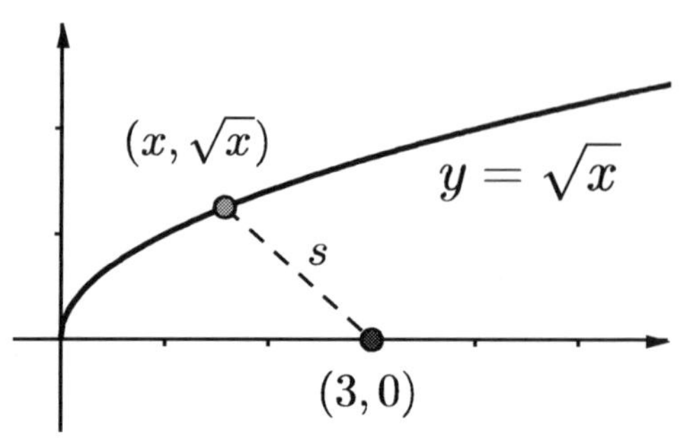

To minimize s, we'll search for hill/valley candidates: places, that is, where its derivative is zero. Since

$$\frac{ds}{dx} = \frac{2(x-3)+1}{2\sqrt{(x-3)^2 + x}},$$

which is zero only when $x = 5/2$ (as you should verify), this value of x must be the one we want. (In this problem, it is geometrically obvious that some point must minimize the distance, but that none will maximize it.) Hence, the point on the curve closest to $(3,0)$ is $(5/2, \sqrt{5/2})$. ♦

Example 3. A farmer has 60 watermelon trees, each of which produces 800 melons per year.* Each additional tree he plants will reduce the output of each tree in his orchard by 10 melons per year. What is the maximum number of melons his orchard can produce in a year?

Solution. Let x be the number of additional trees. Then, with $(60 + x)$ trees in the orchard, each of which produces $(800 - 10x)$ melons per year, the total number of melons per year will be

$$M(x) = (60 + x)(800 - 10x) = -10x^2 + 200x + 48{,}000.$$

Now, $M'(x) = -20x + 200$, which is zero when $x = 10$. Hence, to maximize his yield, the farmer should plant 10 more trees. His orchard will then produce $M(10) = 49{,}000$ melons per year. ♦

Finally, I'll offer some fatherly advice regarding the optimization problems that follow. Do not rush blindly into computations; be sure you understand the problem thoroughly before you compute anything. Read the problem through several times, and draw careful figures whenever geometry is involved. Express the quantity you wish to optimize in terms of your figure's labels (or, if you have no figures, in terms of variables you've introduced and explicitly defined). If this expression involves multiple independent variables, you'll need to eliminate all but one of them before you can use the *single*-variable differential calculus you've learned in this course. Now go forth and optimize.

* Watermelons, of course, grow on vines. But not in this problem.

Exercises.

10. In example 3 above, explain how you can be sure (without using a graphing program) that planting 10 more trees will in fact *maximize* the yield, rather than minimize it. Now explain it a second way.

11. Mack the Finger has 500 feet of fencing with which to enclose (part of) a field. The enclosed area must be a rectangle. A huge building conveniently located on one side of the field will serve as one of the rectangle's sides. Thus, the enclosed area will require only three fenced sides. Find the dimensions of the largest area that can be enclosed this way.

12. Prove that among all possible rectangles with a given perimeter, the one with the greatest area is the square.

13. We can write 10 as the sum of two positive reals x and y in many different ways. [e.g. $4+6$, $3.81+6.19$] Find the way that minimizes the quantity x^3+y^2.

14. A manufacturer who must make a cylindrical can holding 3 liters of liquid wishes to minimize the amount of material used to make the can. Find the dimensions of his ideal can. [*Hint for Americans:* $1L = 1000\ cm^3$.]

15. Louis the King wants to cut a wire into two pieces, one of which he'll bend into a square, the other of which he'll bend into an equilateral triangle. Where should he cut the wire (which is 2 feet long) so that the total area enclosed by the two shapes will be minimized? What should he do if he wishes to maximize the area?

16. Glasses no longer suffice for her purposes, so Nancy decides to build a vodka trough. She'll begin with a rectangular piece of sheet metal: 120 cm long, 20 cm wide. Indulging her mania for symmetry, she'll bend it into three equal parts (as shown at right) to produce the trough's basic shape. To complete her work, she'll need to solder trapezoidal pieces to either side – an easy job for a talented girl like Nancy. Find the angle θ that maximizes the amount of vodka her trough will hold.

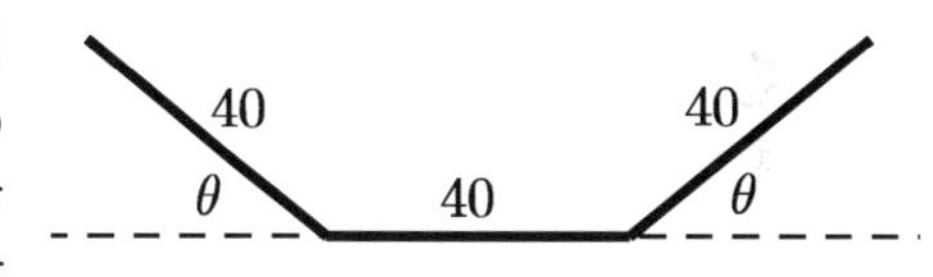

17. Many different cones can be inscribed in the same sphere. Naturally, some cones take up a larger percentage of the sphere than others. To the nearest tenth of a percent, what is the *maximum* percentage of a sphere's volume an inscribed cone can take up?

18. The figure at right represents a sugar cube, each edge of which is 1 unit long. If an ant at point A wants to walk to B, find the length of the shortest path that will take him there. [*Hint: There is a long way and a short way to solve this. Try to minimize not only the length of the ant's walk, but the length of your solution too*.]

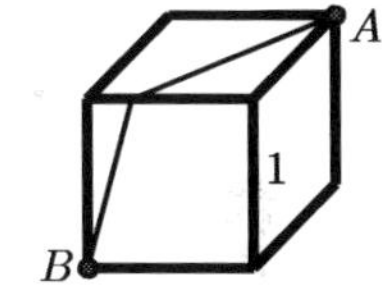

19. A small-time dealer buys high-quality cocaine from his supplier at \$50 per gram. If he charges his customers \$70 per gram, he can sell 100 grams. The more he charges, the fewer grams he'll sell. In fact, his meticulous research indicates that for each dollar he raises his selling price, he'll sell two grams fewer.

a) What should he charge to maximize his profits? (Assume he buys exactly as many grams as he sells.)

b) During a period in which the police are cracking down on drug trafficking, our dealer has fewer competitors, and figures that under these circumstances, each dollar by which he raises his price per gram will result in a loss of only 1 gram of sales. Now how much should he charge?

20. Let $f(x) = x^{2/3}$. Since $f(0) = 0$ and $f(x) > 0$ for all nonzero x, the function clearly attains a *minimum* value. However, this minimum flies under the derivative's radar: $f'(x) = 0$ has no solution. Your problems: First, explain how we know that $f(x)$ is always positive when $x \neq 0$. Second, explain *why* the function's minimum eludes the derivative. [*Hint*: *f is the square of the cube-root function. Use this fact to graph it by hand. Thinking about concavity may help you.*]

Light on Light: Optics and Optimization

"And as he came, he saw that it was Spring,
A time abhorrent to the nihilist
Or searcher for the fecund minimum."
- Wallace Stevens, "The Comedian as the Letter C"

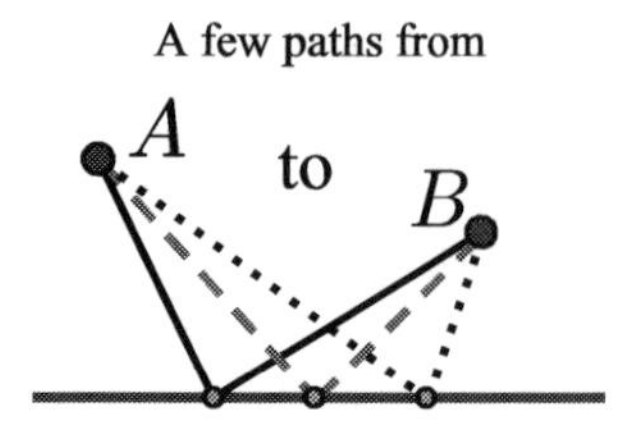

Light, says Heron, chooses the shortest one.

In Chapter 1, I mentioned the **law of reflection**: When a light ray *reflects* off of a surface, it invariably "bounces off" at an angle equal to that at which it struck the surface. This property of light has been known for millennia. Writing around 300 BC, Euclid mentions it casually in his *Optics*, as something already well-known to his audience. Still, there is knowing and then there is *knowing*. In the first century AD, Heron of Alexandria (our Hero) proposed to explain *why* the reflection law holds: Light is lazy.* If we observe a ray of light emanate from point A, reflect off of a surface, and continue to point B, then, claimed Heron, we may be quite sure that the particular two-stage path the ray took was the *shortest* of all such paths from A to B.

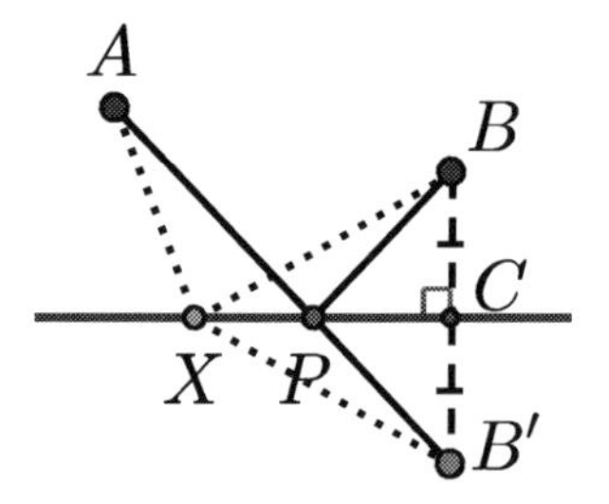

This was no idle accusation. Heron proved his case with a miniature masterpiece of geometric reasoning. Let X be a variable point on the horizontal line in the figure. Light, says Heron, wants to minimize the distance $AX + XB$. This, he claimed, happens precisely when X is at the point P where line AB' crosses the blue line. (B' itself is the reflection of B over the horizontal line.) To justify his claim, Heron noted that for any position X *other than* P, we have

$$\begin{aligned} AP + PB &= AP + PB' && \text{(since } \Delta PBC \cong PB'C\text{).} \\ &= AB' \\ &< AX + XB' && \text{(the shortest distance between two points is a straight line)} \\ &= AX + XB && \text{(since } \Delta XBC \cong XB'C\text{).} \end{aligned}$$

Having thus shown that $AP + PB < AX + XB$ for all positions of X other than P, we conclude that the distance is indeed minimized when X is *at* P. Moreover, the angles of incidence and reflection are equal at P, since both equal $B'\hat{P}C$ (be sure you see why). Thus Heron was correct: The point that minimizes the distance travelled by the light ray is also the one at which these angles are equal.

In one sense, Heron's demonstration "explains" the reflection law, but in another, it only deepens its mystery. Very well, light reflects as it does because Nature seeks the shortest path, but why should Nature seek the shortest path? It's not as if light grows tired. Perhaps the gods desire simplicity in physical law? Lest we get carried too far away, I hasten to note that light does *not* take the shortest path between points when it is *refracted*.

---

* You may know "Heron's formula" from geometry, which expresses a triangle's area in terms of its sides. Heron, confusingly, is often called *Hero*. Neither bird nor warrior, the man wrote his own name Ἥρων, which would have been pronounced something like "Hay-rone" (rhyming with "pay phone"). Transliterated into English, ancient Greek names are often shorn of their final *nu*. The most prominent victim of "nu-dropping" is Πλάτων, known to the non-English-speaking world as Platon, but to us as *Plato*. Since Plato is one of the crucial figures of Western civilization, a consensus on his name's transliteration into English was inevitable. Hero/Heron, brilliant though he was, is known only to a few – and the few cannot make up their minds.

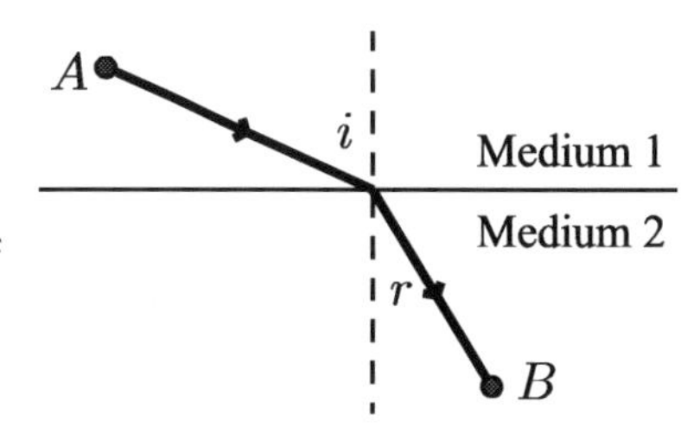

A light ray is *refracted* when it passes from one medium into another (from air to water, for instance, or from water to glass). If the light leaving this page is refracted through your eyeglasses or contact lenses before entering your pupil, you should thank the discoverers of the law of refraction, who made such technology possible. This law is usually called Snell's Law, after the Dutch astronomer who discovered it in 1621. Snell's discovery was, unbeknownst to him, a rediscovery. Thomas Harriot knew of it decades before Snell, but never published the result. Only recently has the law's (presumed) first discoverer come to light: Abu Sa'd al-'Ala' ibn Sahl, who described it in the year 984 in his *On Burning Mirrors and Lenses*.

The oft-discovered **law of refraction** states that

$$\frac{\sin(i)}{v_1} = \frac{\sin(r)}{v_2},$$

where i and r are the angles indicated in the figure above, and v_1 and v_2 are the speeds of light in mediums 1 and 2 respectively. That the refraction law took longer to discover than the reflection law isn't surprising; it involves trigonometry and the idea that light moves at a finite speed. But why does it hold?

In the early 1660s, Pierre de Fermat argued that nature wants to minimize not distance, but *time*. When speed is constant (as in the case of reflection), minimizing distance and time are clearly equivalent. However, when speed can vary, as in the case of refraction, this equivalence no longer holds. To minimize the *length* of its journey from A to B, light would simply travel on line segment AB. But to minimize the *duration* of its journey, light must deviate from this straight-line path. Fermat showed how as follows.

In the figure below, O is a variable point that can side freely on the boundary line between the two media, and l is the fixed horizontal distance between points A and B. Since time is distance over speed, the time light requires to traverse path AOB is

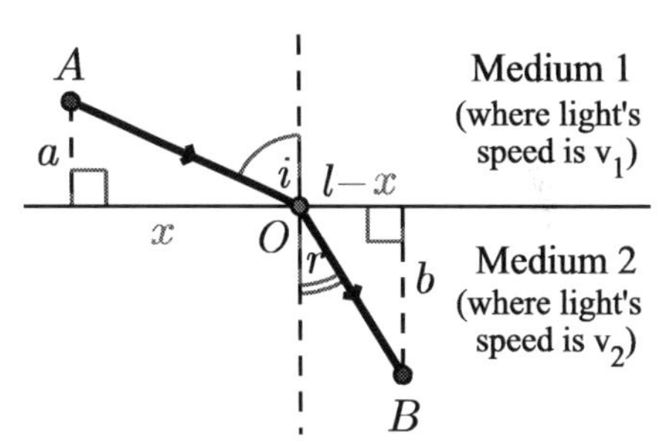

$$t = \frac{AO}{v_1} + \frac{OB}{v_2} = \frac{\sqrt{a^2+x^2}}{v_1} + \frac{\sqrt{b^2+(l-x)^2}}{v_2}.$$

To *minimize* this time, we take the derivative, yielding

$$\frac{dt}{dx} = \frac{x}{v_1\sqrt{a^2+x^2}} - \frac{l-x}{v_2\sqrt{b^2+(l-x)^2}}.$$

Setting this equal to zero, we obtain

$$\frac{x}{v_1\sqrt{a^2+x^2}} = \frac{l-x}{v_2\sqrt{b^2+(l-x)^2}}.$$

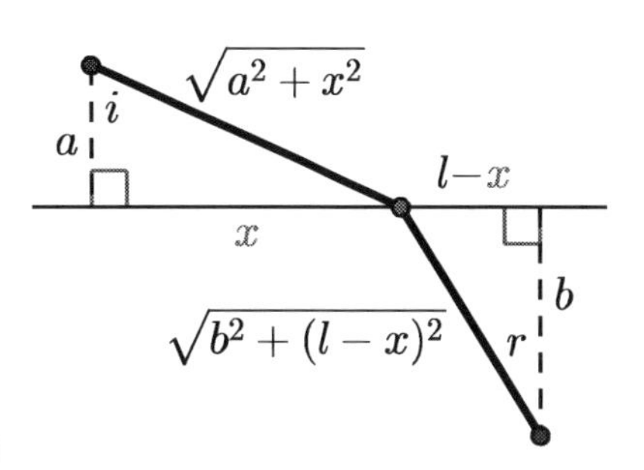

Comparing this to the figure at right (the fruit of basic geometry), we see that the previous equation is equivalent to

$$\frac{\sin(i)}{v_1} = \frac{\sin(r)}{v_2}.$$

Hence, if Nature wishes to minimize the time that a light ray spends travelling, the refraction law must hold!

Fermat's "least-time principle" thus explains the laws of reflection *and* refraction as consequences of Nature's "desire" to minimize the time it takes for light to pass from one point to another.

To attribute desire to Mother Nature and to explain physical laws as the fulfillment of that desire is, at least in recent centuries, a surprising philosophical move. Contemporary physicists, leery of metaphysics, rarely make such assertions. They are content to note agnostically that certain optical principles are inevitable mathematical consequences of the least-time principle, and leave it at that. And yet, one wonders – why should Nature care to optimize? For optimize she certainly does. Optimization is a pervasive theme throughout physics. A freely floating soap bubble, to take an easily understood example, is spherical because a sphere has the *smallest* surface area of all possible surfaces that could contain the bubble's trapped air. Even when multiple bubbles clump together and cling to a flat surface, they do so in forms that solve optimization problems. Perhaps the most remarkable instance of nature's parsimony is revealed by the so-called "principle of least *action*" in mechanics. This principle allows one to explain the path taken by a moving object (say, a thrown ball) *not* by analyzing the forces acting on the object (as one would in Newtonian mechanics), but instead by considering all possible paths the object *might* take to get from one point to another, associating each path with a certain real number called the "action" of that path, and then determining the path of *least* action. Amazingly, the path of least action turns out to be the path actually taken by the object.*

The world is a mysterious place.

---

* Incidentally, a path's *action* is defined in terms of calculus: First we mentally disintegrate the curved path into infinitely many infinitesimally small (hence straight) paths; next, we express the action of each infinitesimal pathlet in terms of the object's kinetic and potential energy (these energies are constant for the object while on the infinitesimal path, where they have "no room to vary"); finally, we sum up (i.e. integrate) these infinitesimal actions to obtain the full path's total action.

Implicit Differentiation

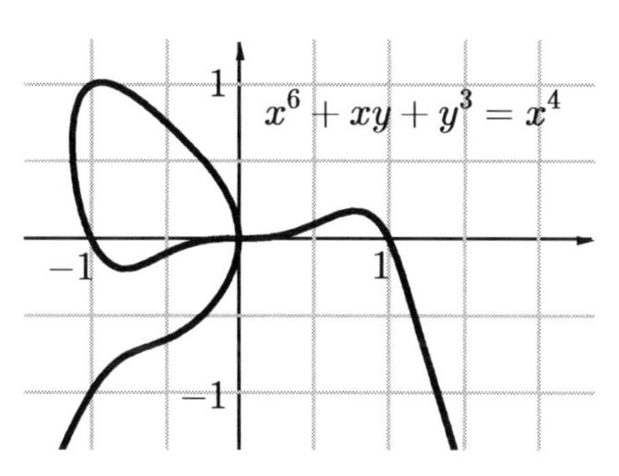

We cannot rewrite $x^6 + xy + y^3 = x^4$ to make y a function of x, for some x-values (e.g. $x = -1$) have *three* corresponding y-values, as the graph at right shows. This will not do; a function can have only one output for each input.

If pressed to find this graph's slope at $(-1,0)$, we might try writing dy/dx as a function of x (and then evaluating it at $x = -1$), but this strategy would fail; our inability to write y as a function of x means that we simply cannot write dy/dx as a function of x.

Happily, we can circumvent this impasse by writing dy/dx as a function of *two* variables: x and y. In this form, the derivative will give us the slope at a point on the curve if we insert *both* of its coordinates. To express the derivative as a function of x and y requires a clever trick known as "implicit differentiation". The basic idea is to *think of* y as a function of x (even if no such function exists!) just long enough to get dy/dx (which *does* exist – albeit as a function of two variables) into the picture.* The details will be made clear in the following example.

Example 1. Find the slope of the graph above at $(-1,0)$.

Solution. The curve's equation is $x^6 + xy + y^3 = x^4$. As discussed above, we'll find dy/dx as a function of two variables by *thinking of* y as some function $[y(x)]$, even though we can't explicitly write down a formula for it. Thinking this way, we rewrite the curve's equation as

$$x^6 + x[y(x)] + [y(x)]^3 = x^4,$$

replacing each y by $[y(x)]$ to remind us that we are thinking of y as a function of x. Now we take the derivative of each side of the equation with respect to x:

$$\frac{d}{dx}(x^6 + x[y(x)] + [y(x)]^3) = \frac{d}{dx}(x^4).$$

Taking care to use the chain rule (and product rule) where needed, this becomes

$$6x^5 + \left([y(x)] + x\frac{dy}{dx}\right) + 3[y(x)]^2\frac{dy}{dx} = 4x^3.$$

Having successfully lured dy/dx onto the scene, we'll collapse our $[y(x)]$'s back into y's:

$$6x^5 + y + x\frac{dy}{dx} + 3y^2\frac{dy}{dx} = 4x^3.$$

---

* This isn't as crazy as it sounds. Consider almost any point on almost any curve; consider, for example, (-1,0) on the curve above. Imagine the curve on a computer screen. Mentally zoom in on the point until no violations of the "vertical line test" are visible. You'll then be looking at the graph of a *function* – whose domain and range have been restricted to those in our viewing window. To "think of y as an implicit function of x" is simply a matter of thinking *locally*: mentally restricting ourselves to a window small enough to ensure that each x-value corresponds to just one y-value.

Standard differential calculus caveat: The preceding analysis can fail at isolated points where the curve *lacks a tangent line*, such as at corners, or, more interestingly, at points where the curve crosses itself. It can also fail where the curve has a *vertical tangent line*, since a vertical line is not, of course, the graph of a *function*, no matter how much we "restrict the window". Note that the curve above has two such points: its self-intersection at the origin, and its loop's leftmost point.

This equation relates x, y, and dy/dx. Solving it for dy/dx will yield the function of two variables we want. As you should verify by doing the algebra, this is

$$\frac{dy}{dx} = \frac{4x^3 - y - 6x^5}{x + 3y^2}.$$

Having expressed dy/dx as a function of x and y, mere arithmetic reveals that $dy/dx = -2$ when $x = -1$ and $y = 0$. Hence, the curve's slope at $(-1,0)$ is $-\mathbf{2}$. ♦

Once you've done one or two implicit differentiation problems, you can dispense with those awkward $[y(x)]$ training wheels; you need only *bear in mind* that y is to be thought of as a function of x. Hence, the process of implicit differentiation might involve computations such as

$$\frac{d}{dx}(y^4) = 4y^3\left(\frac{dy}{dx}\right), \qquad \frac{d}{dx}(\sin y) = (\cos y)\left(\frac{dy}{dx}\right), \qquad \text{or} \qquad \frac{d}{dx}(xy^2) = y^2 + x\left(2y\frac{dy}{dx}\right).$$

Here is an example done in this style.

Example 2. Given the relationship $y \sin y = x + x^4$, express dy/dx as a function of x and y.
Solution. To introduce dy/dx, we'll need to take the derivative of both sides with respect to x:

$$\frac{d}{dx}(y \sin y) = \frac{d}{dx}(x + x^4).$$

Remembering that y is to be thought of as a function of x, the product rule applies on the left-hand side, so this becomes

$$\left(\frac{d}{dx}(y)\right)\sin y + y\frac{d}{dx}(\sin y) = 1 + 4x^3$$

Continuing, this becomes – note the chain rule in the left-hand side's second term! –

$$\left(\frac{dy}{dx}\right)\sin y + y(\cos y)\left(\frac{dy}{dx}\right) = 1 + 4x^3.$$

Solving for dy/dx yields

$$\frac{dy}{dx} = \frac{1 + 4x^3}{\sin y + y \cos y}. \quad ♦$$

According to the "power rule", we obtain a power function's derivative by reducing its power by 1 and multiplying by the old power. Initially, we proved that this statement holds for any whole number power. Bit by bit, we proved that the power rule holds for all rational powers. We can now finish the story.

Claim. The power rule holds for *all real powers*.
Proof. Let $y = x^r$, where r is any real number. Taking logarithms of both sides yields $\ln y = r \ln x$. After implicit differentiation, this becomes $(1/y)\,(dy/dx) = r/x$. Finally, some algebraic massage:

$$\frac{dy}{dx} = \frac{ry}{x} = \frac{rx^r}{x} = rx^{r-1}.$$

Thus, the power rule holds for all real numbers, as claimed. ■

Exercises.

21. In each of the following problems, find dy/dx as a function of x and y.

a) $x^3 + y^3 = 3x$ b) $x + x^4 = y - y^5$ c) $\tan y + x \sin x = y$ d) $x^5 - y^3 = 3xy$

e) $e^{x+5y} = y^2$ f) $\sin(x \cos y) = y$ g) $3\arctan y = \ln(x + y)$ h) $\sqrt{xy} + 4 = y$

22. Does the graph of $x^3 + y^3 = 3x$ have a horizontal tangent line at any of its points? If so, at which one(s)?

23. Find the points where the graph of $x + x^4 = y - y^5$ crosses the y-axis, and the curve's slope at each such point.

24. Repeat the previous exercise, changing "y-axis" to "x-axis."

25. Find the slope of the graph of $y \arctan y = \ln(x + y)$ at the point where it crosses the x-axis.

26. a) If $x^2 + 2x + y^2 - 4y + 6 = 0$, then what is dy/dx (in terms of x and y)?

b) Feed your favorite graphing program the preceding equation, and it will return a bare set of coordinate axes. Explain why it does this.

c) Savor the exquisitely *precise* sort of nonsense we obtain by combining parts (a) and (b). Surely this would have appealed to Lewis Carroll, who was himself a mathematician.

27. a) Find the points at which the graphs of $x^2 + 3y^2 = 12$ (an ellipse) and $9x^2 - 3y^2 = 18$ (a hyperbola) cross.

b) Pick any one of the points you found in part (a) and prove that the curves cross at right angles there.

Related Rates

"The times they are a-changin'."

- Bob Dylan

Change begets change. As you blow air into a balloon, its volume increases, its surface area increases, the thickness of its rubber decreases, and so forth. These parameters change at various (and varying) *rates*, which themselves are related in various ways. Solving a few "related rates" problems is a calculus rite of passage. No new theory is involved, so I'll simply offer a few examples. In the solution to the first, I provide some hints for how to approach problems of this sort, so do read with care.

Example 1. A 13-foot ladder leans against a wall. Suppose its bottom slides away from the wall at a constant rate of $1/10$ of a foot per second. How fast is the top sliding down the wall at the moment when it is 1 foot above the ground?

Solution. We begin by drawing a figure and defining variables. Note well: Our figure depicts the general situation, *not* the particular moment when the ladder's top is 1 foot above the ground. **We must initially leave some "slack" in our model if it is to be useful.** We are, after all, interested in how y is *changing* when $y = 1$. For something to *change*, it cannot be frozen; we must allow for infinitesimal changes if we are to use calculus.

Next, in terms of the variables we've introduced, we'll express what we know and what we *want* to know. If t is the number of seconds since the ladder began moving, then

We know: $\frac{dx}{dt} = \frac{1}{10}$ ft/sec.

We want: $\frac{dy}{dt}$ when $y = 1$.

We know something about one derivative, and want to know something about another. **To relate two variables' *derivatives*, we must relate the variables themselves.** (That is, we must find an equation containing both variables.) Sometimes the variables' relationship will be given, sometimes it will be obvious, and sometimes you'll have to fight for it. In the present problem, the relationship is obvious: x and y are linked by the Pythagorean theorem, which tells us that

$$x^2 + y^2 = 169$$

for all positions of the ladder.

Having related our variables, we now relate their derivatives *by implicitly differentiating both sides with respect to* t. (Remember: x and y are themselves unspecified functions of t.)

$$\frac{d}{dt}(x^2 + y^2) = \frac{d}{dt}(169),$$

$$2x\frac{dx}{dt} + 2y\frac{dy}{dt} = 0.^{*} \qquad \bigstar$$

This equation relates the four quantities x, y, dx/dt, and dy/dt at any point in time (and hence, for any position of the ladder). Thus, if we know three of these quantities at a particular moment, we can use equation ($\bigstar$) to find the value of the fourth at that same moment.

In particular, we can achieve our goal for this problem of finding dy/dt when $y = 1$ if we can find the three other values *at that same moment*. Two were given to us: $y = 1$, and $dx/dt = 1/10$. To find x, we draw a figure representing the moment in question. The Pythagorean Theorem reveals that at that moment, $x = \sqrt{168}$. Substituting our three values into ($\bigstar$), we find that **when $y = 1$**,

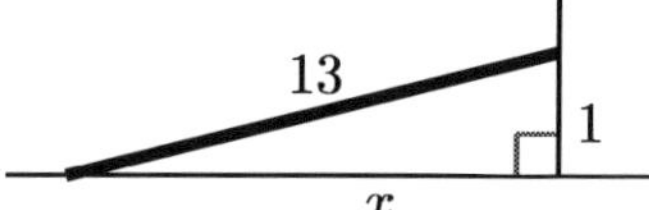

$$2\sqrt{168}\left(\frac{1}{10}\right) + 2(1)\frac{dy}{dt} = 0\,.$$

Or equivalently, $\frac{dy}{dt} = -\frac{\sqrt{42}}{5}$ when $y = 1$, as you should verify.

In conclusion, when the ladder's end is one foot above the ground, its height is decreasing at $\sqrt{42}/5 \approx 1.30$ feet per second. ♦

Before we consider a second example, take stock of the strategy we used in the first. First, we drew a picture of the general situation (taking care not to "freeze" it too early), introduced some variables, and wrote down – in terms of our variables' derivatives – what we knew and what we wanted to know. Second, we found a relationship between the variables themselves, and took derivatives of both of its sides, using implicit differentiation. This yielded a new equation relating the variables *and their derivatives*. Third, by applying this new equation to the particular moment with which the problem was ultimately concerned, we solved the problem.

This same basic strategy will guide you through most related rates problems.

---

* Note the chain rule! Were we to do this with training wheels, we'd write $[x(t)]$ in place of x, and note that, for example,

$$\frac{d}{dt}(x^2) = \frac{d}{dt}([x(t)]^2) = 2[x(t)]\frac{dx}{dt} = 2x\frac{dx}{dt}.$$

The second equals sign is justified by the chain rule.

Example 2. A huge tank shaped like an inverted square-based pyramid is 5 m high. The sides of its "base" are 2 m long. Water pours into the tank at a rate of 3 m^3/min. How rapidly is the depth of the water in the tank (at its deepest point) rising when the water in the tank is 4 m deep?

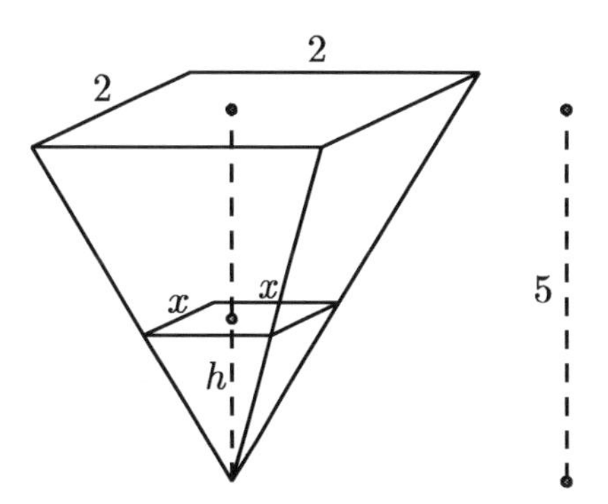

Solution. The figure at right represents the situation t seconds after the water begins pouring in, where h is the water's depth at time t. If we let V represent the volume of water in the tank at time t, we are *given* that $dV/dt = 3$. We wish to find dh/dt when $h = 4$.

To relate dh/dt and dV/dt, we'll first relate h and V. Recall Democritus's famous formula for a pyramid's volume: $1/3$ of the prism with the same base and height.* The prism containing the watery pyramid is a rectangular box of base area x^2 and height h, so Democritus tells us that $V = (1/3)x^2h$. This is a good start, but it has one variable too many. Before we take derivatives, we'll need to eliminate x. This is easily done. Since the watery pyramid is *similar* to the tank itself, we have $x/h = 2/5$. Hence, $x = (2/5)h$. Substituting this into our volume formula yields

$$V = \frac{4}{75}h^3,$$

our long-sought relationship between V and h.

Differentiating both sides with respect to t (and remembering the chain rule!), we obtain

$$\frac{dV}{dt} = \frac{4}{75}(3h^2)\frac{dh}{dt}.$$

At the moment when $h = 4$, we know that $dV/dt = 3$. Substituting these two values into the previous equation, we find that $dh/dt = 75/64$ at this same moment.

Thus, when the water is 4 m deep, its height is rising at $75/64$ meters per minute. ♦

Example 3. In physics, an object's *kinetic energy* is half its mass times the square of its velocity. (In the MKS system, the unit of energy is the *joule*.) Suppose a 4 kg object is accelerating at a constant rate of 10 m/s^2. When its velocity is 20 m/s, at what rate is its kinetic energy changing?

Solution. There's no picture to draw in this case. Let v and K be the object's velocity and kinetic energy at time t. We are given that $dv/dt = 10$, and we want dK/dt when $v = 20$.

Translating the problem's first sentence into symbols (with a 4 kg mass) tells us that $K = 2v^2$. Taking derivatives of both sides with respect to t, we obtain

$$\frac{dK}{dt} = 4v\frac{dv}{dt}.$$

At the moment when $v = 20$, this becomes $dK/dt = 4 \cdot 20 \cdot 10 = 800$, so when the object's velocity is 20 m/s, its kinetic energy is increasing at 800 joules per second. ♦

* This result may have been unfamiliar to you a moment ago, but it isn't now. Commit it to memory.

Exercises.

28. St. Peter places an hourglass before you. "Mark ye how the sands fall into the lower chamber," he instructs, "Yea, even as souls fall into perdition. The heap of fallen grains is conical, and lo, its height always equaleth the radius of its base. Observe the steady rate at which the sands fall: five cubic cubits every minute." Frightened, but fascinated, you look, and verily, it is so. "Ere these sands run out," Peter resumes, with a grave look, "tell me, when the sandy cone is 2 cubits high, how rapidly doth its height increase?"

29. A ten-foot ladder leans against a wall. If its top slides down the wall at a constant rate of 2 feet per second, how fast is the acute angle between ladder and ground changing when the ladder's top is 5 feet above the ground? (And why are ladders in calculus story problems always placed with so little regard for safety?)

30. Air is blown into a perfectly spherical balloon at a constant rate of 3 cubic inches per second. When its radius is half an inch, how quickly is the balloon's surface area increasing?

[*Hint: Have you* ***still*** *not memorized the volume and surface area formulas for a sphere? They're not hard to recall. Start with volume. Obviously, it has a factor of π (since spheres are "circular") and a factor of r^3 (since it concerns volume). You need only recall the remaining factor of $4/3$. Surface area is even easier: it is exactly four times the area of the sphere's equatorial circle! (Or, derive it by taking the volume formula's derivative with respect to r.)*]

31. Boyle's Law states that if we keep a fixed quantity of gas at a constant temperature, the product of its volume V and the pressure P it exerts on its container will remain constant. (So, for example, compressing the gas into a space of half its original volume will double the pressure it exerts on the tank's walls.)

a) Suppose we hold some gas at a constant temperature in a tank, the volume of which we can change by pushing in (or drawing out) a piston. From Boyle's law, derive an equation that relates volume, pressure, rate of change of pressure, and rate of change of volume.

b) Suppose that the gas initially takes up 1000 in^3 and exerts a pressure on the tank's walls of 50 psi (pounds per square inch). If the tank's volume is then compressed at a constant rate of 50 in^3/min, at what rate is the pressure initially *increasing*?

c) At the moment when the piston has reduced the tank's volume to one fifth of its initial volume, how rapidly is the pressure increasing?

32. A passing spherical balloon released by Robert Boyle knocks a conical pile of sand from an unstable ladder into a pond, producing a series of concentric circular ripples. The outermost ripple increases its radius at a rate of 2 meters per second. Calculate the rate at which the area contained by the outermost ripple is changing when the radius of this ripple is 4 meters.

33. A lighthouse beacon on an island 2 miles from a straight coast revolves 10 times per minute at a constant speed. The illuminated point where the beacon's beam meets the coast sweeps down the coast once per revolution.

a) Convince yourself that the illuminated point does *not* move at a constant speed.

b) Let P be the point on the coast nearest the beacon. When the illuminated point shines on a location two miles down the shore from P, how fast is it moving?

34. At a constant rate of 4 feet per second, a 6-foot tall streetwalker (6 feet tall in her stiletto heels, that is) walks away from a 14-foot high streetlight.

a) How fast is the tip of her shadow moving (relative to the fixed streetlight)?

b) How fast is her shadow lengthening? [*Hint: Bear in mind that as she walks, both ends of her shadow move.*]

Chapter π
Limits

A Different Dialect: Limits

In this book's first pages, we used infinitesimals to establish the formula for a circle's area. Here is the same basic argument, but now stripped of infinitesimals and re-expressed in terms of *limits*.

Claim. The area of a circle with radius r is πr^2.

Proof. Inscribe a regular n-gon in the circle. The greater n is, the better the polygon approximates the circle. In fact, by making n sufficiently large, we can bring the polygon's area as close to the circle's as we wish. In other words, *the circle's area is the* ***limit*** *of the n-gon's area as n approaches infinity.* Or, expressed in symbols,

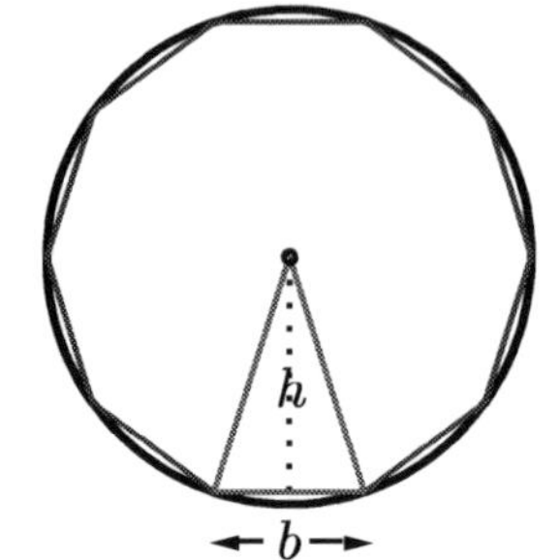

$$A_{circle} = \lim_{n\to\infty} A_{n-gon} \,.$$

Our regular n-gon's area is n times that of the triangle shown in the figure, so $A_{n-gon} = n(bh/2) = Ph/2$, where P is the polygon's perimeter. Thus,

$$\begin{aligned} A_{circle} &= \lim_{n\to\infty} A_{n-gon} \\ &= \lim_{n\to\infty} (Ph/2). \end{aligned}$$

Now, to actually "take the limit", we imagine what happens to each part of the expression as n approaches infinity: Clearly, P gets closer and closer to $2\pi r$ (the circle's circumference), h approaches r (the circle's radius), and 2, being a constant, remains unchanged. Consequently,

$$\begin{aligned} A_{circle} &= \lim_{n\to\infty} A_{n-gon} \\ &= \lim_{n\to\infty} (Ph/2) \\ &= (2\pi r)r/2 \\ &= \pi r^2. \qquad \blacksquare \end{aligned}$$

Infinitesimals and limits are two dialects of the same language. If you know one, you can speak the other by making a few little changes. For example, we can regard a derivative as a **ratio of *infinitesimal* changes**, or, switching dialects, as **the *limit* of a ratio of small *real* changes**. In the latter case, we imagine increasing a function's input by a small *real* amount (Δx), finding the corresponding small real change in output (Δy), taking the ratio of these changes, and finally, *taking the limit* as the input change Δx approaches zero (i.e. as it becomes more and more like an infinitesimal). Thus, if y is a function of x, we have

$$\boxed{\frac{dy}{dx} = \lim_{\Delta x\to 0} \frac{\Delta y}{\Delta x}.}$$

In a limit-based calculus course, the boxed equation is given (on the first day) as the derivative's *definition*. Over the following days and weeks, one uses this "**limit definition of the derivative**" first to compute some derivatives directly (as in the following example), then to derive the familiar rules of the calculus.*

* Since the expression $\Delta y/\Delta x$ is a quotient of differences, it is often called "the difference quotient". A derivative can therefore be described as *the limit of the difference quotient* (as its denominator vanishes).

Example. Using the *limit definition* of the derivative, show that if $y = x^2$, then $\frac{dy}{dx} = 2x$.

Solution. If we increase the function's input from x to $x + \Delta x$ (where Δx is a small *real* number), its output changes from x^2 to $(x + \Delta x)^2$. Hence, the limit definition for derivatives yields

$$\frac{dy}{dx} = \lim_{\Delta x \to 0} \frac{\Delta y}{\Delta x} = \lim_{\Delta x \to 0} \frac{(x + \Delta x)^2 - x^2}{\Delta x} = \lim_{\Delta x \to 0} \frac{x^2 + 2x\Delta x + (\Delta x)^2 - x^2}{\Delta x} = \lim_{\Delta x \to 0} (2x + \Delta x) = 2x\,. \ \blacklozenge$$

Note how this limit-based view of derivatives stays entirely within the real number system. Some find the absence of infinitesimals comforting, others find it a bit dull. There is merit in both viewpoints.

Last observations: *Limits need not exist.* For example, as x approaches infinity, $\cos x$ doesn't approach any specific value; it just cycles over and over again through all of the real numbers between -1 and 1. Other functions have values that, in the limit, grow without bound. (More colloquially, they "blow up".) In such cases, we say that the limit is infinity. For example, we would write that $\lim_{x \to \infty} x^2 = \infty$.

Exercises.

1. Use (as in the preceding example) the limit definition of the derivative to find the derivatives of the following:.

a) $y = x^2 + x$ b) $y = x^3$ c) $y = 1/x$ d) $y = \sqrt{x}$ [*Hint: Rationalize the numerator.*]

2. Explain why the derivative's limit definition can be restated in this form: $f'(x) = \lim_{\Delta x \to 0} \frac{f(x + \Delta x) - f(x)}{\Delta x}$.

3. Use the figure at right to convince yourself that $f'(a) = \lim_{x \to a} \frac{f(x) - f(a)}{x - a}$.

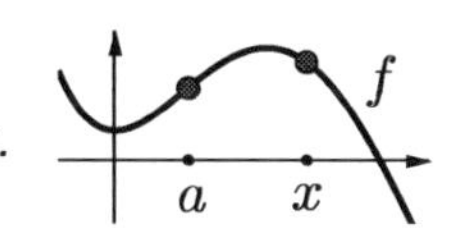

[*Hint: Find an expression for the slope of the line joining the two points on the graph at right. As x approaches a, this "cutting line" approaches the* ***tangent*** *to the graph when x = a.*]

4. All the familiar rules of differential calculus can be established with limit arguments instead of infinitesimals. For example, to prove that the derivative of a sum is the sum of the derivatives, we can note that

$$\begin{aligned}(f(x) + g(x))' &= \lim_{\Delta x \to 0} \frac{(f(x + \Delta x) + g(x + \Delta x)) - (f(x) + g(x))}{\Delta x} \\ &= \lim_{\Delta x \to 0} \frac{(f(x + \Delta x) - f(x)) + (g(x + \Delta x) - g(x))}{\Delta x} \\ &= \lim_{\Delta x \to 0} \left[\frac{f(x + \Delta x) - f(x)}{\Delta x} + \frac{g(x + \Delta x) - g(x)}{\Delta x}\right] = f'(x) + g'(x).\end{aligned}$$

a) Justify each equals sign in the preceding proof. [*Hint for the last one: Recall exercise* 2.]

b) Construct a similar limit-based proof for the derivative's other linearity property: $(cf(x))' = cf'(x)$.

5. By thinking about the functions that follow, find their indicated limits, or state that they do not exist.

a) $\lim_{x \to 1} 2x^2 + 3$ b) $\lim_{x \to 3} \frac{x + 5}{2 - x}$ c) $\lim_{x \to \infty} \sin(x + 2)$ d) $\lim_{x \to \infty} \ln(5x)$

e) $\lim_{x \to \infty} e^{-x}$ f) $\lim_{x \to \infty} \arctan x$ g) $\lim_{x \to 0} \left(\frac{x - 1}{2x^2 + 3x + 4}\right)$ h) $\lim_{x \to -\infty} \left(\frac{3 + 2^x}{\arctan x}\right)$

6. The notation $x \to a^+$ means "x approaches a **from the right** (on the number line)." For example, the values $2.01, 2.001, 2.0001$ (and so forth) approach 2 from the right. Naturally, we could approach a from the left as well, adjusting the notation in the obvious way ($x \to a^-$). With these ideas in mind, find the following limits:

a) $\lim_{x \to 0^+} (1/x)$ b) $\lim_{x \to 0^-} (1/x)$ c) $\lim_{x \to 0} (1/x^2)$ d) $\lim_{x \to \infty} (1/x)$ e) $\lim_{x \to 0^+} \ln x$ f) $\lim_{x \to \frac{\pi}{2}^+} (\tan x)$ g) $\lim_{x \to 1^-} \left(\frac{x + 1}{x - 1}\right)$

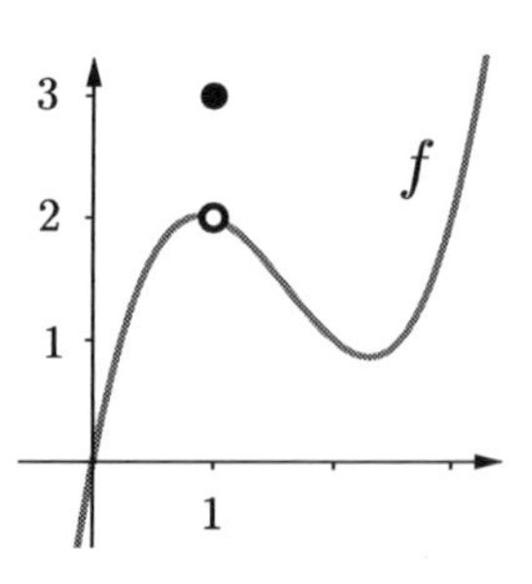

7. Limits describe what happens as we *approach* a point; not what happens *at* the point. This subtle distinction is easily grasped if we read $\lim_{x \to a} f(x) = L$ as meaning

Whenever x is infinitesimally close to a, then $f(x)$ will be infinitesimally close to L.

Consider the graph of the ill-bred function at right, which has a jarring discontinuity at $x = 1$. (Instead of sending 1 to 2 as it "ought to", the function f sends 1 to 3.)

Question: What is $\lim_{x \to 1} f(x)$? [*Hint: Test your answer against the statement above*.]

Zero Over Zero

The expression $5/0$ is undefined because it asks a question ("0 times *what* is 5?") with no valid answer. We also leave $0/0$ undefined, but for quite a different reason: The question it asks ("0 times what is 0?") has too many valid answers! All real numbers fit the bill. But if all numbers were equal to $0/0$, then all numbers would be equal to the same thing, and since things equal to the same thing are themselves equal, *all numbers would be equal*, which would effectively destroy the entire subject of mathematics.

Differential calculus flirts with this abyss. It approaches $0/0$ as closely as possible, and channels its destructive power into more useful directions. Whenever we take a derivative, we approach those twin black holes of $0/0$, slipping away just before we are sucked down. What is a derivative, after all, but a ratio of infinitesimals? A ratio, that is, of two numbers the naked eye can't distinguish from zeros.*

The *limit* of a ratio whose numerator and denominator *both* approach 0 is decided by a tug of war. The numerator "tries" to pull the fraction's value to zero, since zero divided by anything (other than zero) is zero. On the other hand, the denominator tries to pull the fraction's value to infinity, since dividing anything by something "zeroish" (i.e. comparatively miniscule) yields something enormous.†

When a vanishing numerator and a vanishing denominator engage in a tug of war, the outcome is determined by the *rates* at which they vanish. It should therefore come as no surprise that someone has proved a theorem relating such contests to their contestants' *derivatives*. It is called **L'Hôpital's Rule**. As a bonus, L'Hôpital's Rule also covers the case of an ∞/∞ tug of war, in which the numerator tries to pull the ratio up to infinity, and the denominator tries to drag it down to zero.

L'Hôpital's Rule. If $f(x)$ and $g(x)$ both *vanish* (approach 0) or both *blow up* (approach $\pm\infty$) as x approaches a (where a could be either a real number or $\pm\infty$), then

$$\lim_{x \to a} \frac{f(x)}{g(x)} = \lim_{x \to a} \frac{f'(x)}{g'(x)}.$$

Partial Proof. I'll prove the case in which $f(x)$ and $g(x)$ vanish as x approaches a real number a. The other cases can be reduced to this one, but the details are hairy.

* In a characteristically cryptic poem ("More Life – went out – when He went"), Emily Dickinson at one point employs the phrase "an Ampler Zero", which, it seems to me, could double as a poetic description of an infinitesimal.

† Compare: $\lim_{x \to 0} \frac{x^2}{x+5} = 0$ (top vanishes), $\lim_{x \to 5} \frac{x^2}{(x-5)^2} = \infty$ (bottom vanishes), $\lim_{x \to \pi} \frac{\sin x}{(x-\pi)^2} = ?$ (both vanish).

Since we are assuming that $f(x)$ and $g(x)$ *vanish* as x approaches a, we have $f(a) = g(a) = 0$.*
Thus,

$$\lim_{x\to a}\frac{f(x)}{g(x)} = \lim_{x\to a}\frac{f(x)-f(a)}{g(x)-g(a)} = \lim_{x\to a}\frac{\left(\dfrac{f(x)-f(a)}{x-a}\right)}{\left(\dfrac{g(x)-g(a)}{x-a}\right)} = \frac{f'(a)}{g'(a)} = \lim_{x\to a}\frac{f'(x)}{g'(x)}.$$

The first equals sign's validity has already been justified. The second is obvious, as is the fourth – if you read it from right to left. This leaves only the third, which follows from exercise 3. ■

The Marquis de L'Hôpital, Guillaume François Antoine, was a gifted expositor of mathematics (he even wrote the world's first calculus textbook in 1696), but some of his ostensibly original discoveries, including the rule bearing his name, were in fact the work of Johann Bernoulli, with whom he had an odd financial arrangement: L'Hôpital paid Bernoulli for the right to pass off some of Bernoulli's work as his own.

Justly named or not, L'Hôpital's rule is easy to use.

Example 2. Find $\lim\limits_{x\to 0}\dfrac{\sin x}{x}$.

Solution. Since the top and bottom vanish in the limit, L'Hôpital's Rule applies, and tells us that

$$\lim_{x\to 0}\frac{\sin x}{x} = \lim_{x\to 0}\frac{\cos x}{1} = 1\,. \quad \blacklozenge$$

Example 3. Find $\lim\limits_{x\to\infty}\dfrac{x^2}{e^x}$.

Solution. Since both functions blow up in the limit, L'Hôpital's Rule applies. It tells us that

$$\lim_{x\to\infty}\frac{x^2}{e^x} = \lim_{x\to\infty}\frac{2x}{e^x}.$$

We are thus faced with a second ∞/∞ limit. Applying L'Hôpital's Rule again yields

$$\lim_{x\to\infty}\frac{x^2}{e^x} = \lim_{x\to\infty}\frac{2x}{e^x} = \lim_{x\to\infty}\frac{2}{e^x} = 0\,. \quad \blacklozenge$$

Before you call on the good Marquis, you must, of course, verify that his rule applies. The functions in the ratio must either both vanish, or both blow up. Trying to apply L'Hôspital's Rule when this isn't the case is a terrible *faux pas* – like trying to apply the Pythagorean Theorem to an oblique triangle.

Exercises.

7. Asked to evaluate the limit below, Dippy Dan applies L'Hôpital's Rule, and concludes that

$$\lim_{x\to 1}(2/3x) = \lim_{x\to 1}(0/3) = 0\,.$$

Explain why L'Hôpital's Rule does *not* apply here, then find the limit's actual value.

---

* This follows from our blanket assumption in this book that our functions are well-behaved. Here, "well-behaved" specifically means that our functions (and their derivatives) are *continuous*: their graphs can be drawn without lifting the pen from the page (unlike the function in exercise 7). One can prove that if a continuous function's output approaches 0 as its input approaches a, then its output (not surprisingly) must actually be 0 when its input actually is a. Readers intrigued by such logical niceties will want to study *real analysis*.

8. Evaluate the following limits.

a) $\lim\limits_{x\to 1}\dfrac{\ln x}{x-1}$ b) $\lim\limits_{x\to 0}\dfrac{\sin(3x)}{x}$ c) $\lim\limits_{x\to \pi}\dfrac{1+\cos x}{(x-\pi)^2}$ d) $\lim\limits_{x\to 0}\dfrac{x}{e^{2x}-1}$ e) $\lim\limits_{x\to 3}\dfrac{1+2^x}{(x-1)^2}$ f) $\lim\limits_{x\to \pi/2}\dfrac{\sin(2x)}{\cos x}$

9. In the long-run (as $x \to \infty$), a polynomial "looks like" its leading term. Consider $p(x) = -2x^3 + 8x^2 + 14$. For modest values of x (say, $x = 2$), the values of all three of the polynomial's terms will be on the same general order of magnitude (when $x = 2$, their values are $-16, 32$, and 14), so all three contribute to the output's overall "size."

However, when x is large (say, 10^{12}, which is still a far cry from infinity), the basic "size" of p's output will be determined exclusively by its leading term. After all, $p(10^{12}) = -2(10^{36}) + 8(10^{24}) + 14$, and compared to 10^{36}, that final 14 might as well be infinitesimal; moreover, since 10^{24} is only *one trillionth* of 10^{36}, even the middle term is effectively zero. Hence, for large x, we conclude that $p(x) \approx -2x^3$.

We can discern a *rational function*'s "end-behavior" (*i.e.* its limit as x goes to infinity) by applying these ideas to the two polynomials from which it is built. For example,

$$\lim_{x\to\infty}\frac{-2x^3+8x+14}{4x^4-9x^2+2x+1} = \lim_{x\to\infty}\frac{-2x^3}{4x^4} = \lim_{x\to\infty}\frac{-1}{2x} = 0\,.$$

We could have obtained this by applying L'Hôpital's Rule three successive times, but thinking about long-run behavior is far better: It explains *why* the result is what it is, and – once you get the hang of it – it is quicker, too. By thinking about long-run behavior of polynomials (or otherwise), evaluate the following limits.

a) $\lim\limits_{x\to\infty}\dfrac{5x^3+18x-7}{2x^4+x^3+x+100{,}000}$ b) $\lim\limits_{x\to\infty}\dfrac{5x^7+18x-7}{10x^6-x^3+5x}$ c) $\lim\limits_{x\to\infty}\dfrac{18x^9+400x^8+x^2+42x-7}{-6x^9+17x^7+5x^3-19x+1}$

d) $\lim\limits_{x\to\infty}\dfrac{-x^{365}+2x^{52}+x^{12}-2}{-2x^{365}+x^{52}+2x^{12}-2}$ e) $\lim\limits_{x\to 1}\dfrac{8x^3-10x-7}{10x^6-x^3+5}$ f) $\lim\limits_{x\to\infty}\dfrac{8x^3-10x-7}{10x^6-x^3+5}$

10. The rates at which functions "blow up" to infinity (as x approaches infinity) can vary wildly. The squaring function, for example, increases quickly. The logarithm function takes its time.

a) How large must x be for x^2 to equal 1,000,000?

b) With pen and paper alone (no calculator!), get a feel for how large x must be for $\ln x$ to equal 1,000,000. [*Hint: Use two crude, but obvious, inequalities: $e^n > 2^n$ for any whole number n, and $2^{10} > 10^3$. Together with basic exponent properties, these will help you find a lower bound on the number of digits x must have.*]

c) Given that $\ln x$ climbs so slowly to infinity, you should find it obvious (after a bit of thought) that

$$\lim_{x\to\infty}\frac{\ln x}{x^2} = 0 \quad \text{and that} \quad \lim_{x\to\infty}\frac{x^2}{\ln x} = \infty.$$

Think about these until they are obvious to you. Then confirm their truth with L'Hôpital's Rule.

d) Forget about L'Hôpital. By *thinking* about the top and bottom functions, evaluate $\lim\limits_{x\to\infty}\dfrac{\ln x}{5x^9-10x^2}$.

e) Using L'Hôpital's Rule, evaluate $\lim\limits_{x\to\infty}\left(\ln x/\sqrt{x}\right)$ and interpret the result: In a race between these two notorious slowpokes (logarithm and square root), which goes to infinity faster?

11. Power functions blow up rapidly. Exponential functions blow up still faster. Using L'Hôpital's Rule, explain why, regardless of how large n is, the exponential function e^x goes to infinity faster than the power function x^n.

12. Evaluate the following limits by the method of your choice.

a) $\lim\limits_{x\to\infty}\dfrac{x^2}{2^x}$ b) $\lim\limits_{x\to\infty}\dfrac{\ln(\ln x)}{5+\ln x}$ c) $\lim\limits_{x\to\infty}\dfrac{e^x}{\ln x}$ d) $\lim\limits_{x\to\infty}\dfrac{e^{-x}}{2^{-x}}$ e) $\lim\limits_{x\to 0}\dfrac{2x^7-3x+2}{3x^8+2x+6}$ f) $\lim\limits_{x\to\infty}\dfrac{2x^7-3x+2}{3x^8+2x+6}$

Part II
Integral Calculus

Chapter 4
The Basic Ideas

Integrals: Intuition and Notation

"We could, of course, use any notation we want; do not laugh at notations; invent them, they are powerful. In fact, mathematics is, to a large extent, the invention of better notations."

- Richard Feynman, *Lectures on Physics*, Chapter 17, Section 5.

In principle, finding the area of a polygon (a figure with straight boundaries) is easy: We just chop it into triangles, find their areas, and add them up. What could be simpler? Even the village idiot knows the formula for a triangle's area, and many a bright twelve year old can explain *why* it holds, too.*

Finding the area of a region with a *curved* boundary is trickier, since we can't chop it into a finite number of "nice" pieces. (Here, "nice" just means having a shape whose area we can find.) We can, however, chop it into *infinitely* many nice pieces by making each piece *infinitesimally* thin. For example, we can (mentally) chop the shaded region at right into infinitely many infinitesimally thin rectangles, as the schematic figure below suggests. (Naturally, you must imagine *infinitely many* rectangles there, not just the twenty I have actually drawn.) If we can somehow find these rectangles' areas and sum them up, we will have the region's area.

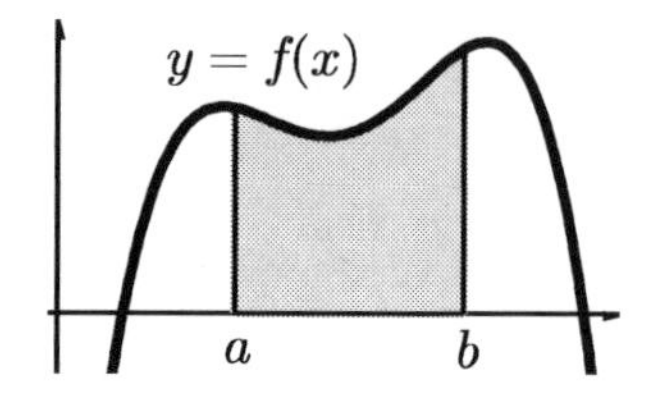

This idea sounds promising in theory, but can we sum infinitely many infinitesimal areas in practice? Thanks to the magic of integral calculus, the answer is *yes*. The magic's deep source, which you will soon behold, is a result called **The Fundamental Theorem of Calculus**. But before you can understand the Fundamental Theorem, you must, as any adept of the dark arts will appreciate, be initiated into the integral calculus's *symbolic* mysteries.

Consider a typical infinitesimal rectangle standing at a typical point, x. Since its height is $f(x)$ and its width is dx (an infinitesimal bit of x), the rectangle's area is $f(x)dx$. The whole region's area is an infinite sum of these $f(x)dx$'s: one for each possible position of x from a to b. To indicate an infinite sum of this sort, we use Leibniz's elegant integral symbol, $\int$, which he intended to suggest a stretched out "S" (for "Sum"). Using this notation, we express the whole region's area as follows:

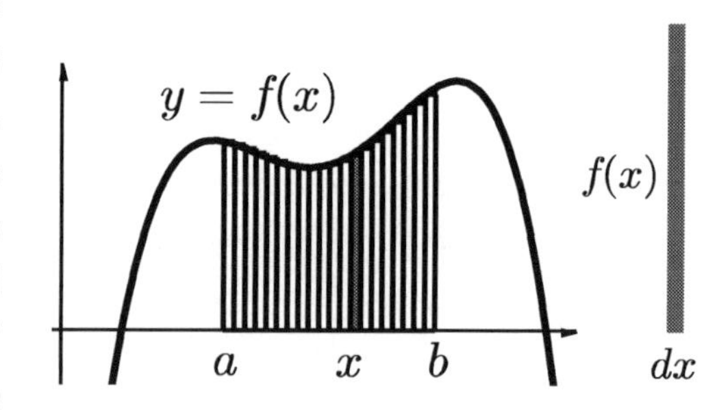

$$\int_a^b f(x)dx.$$

The values a and b near the top and bottom of the integral sign are the **boundaries of integration**, the values of x at which we start and stop summing up the $f(x)dx$'s.

Half the task of learning integral calculus is developing an intuitive feel for this notation and learning to think in terms of infinite sums of infinitesimals. The subject's other half is about *evaluating* integrals – determining their actual numerical values. To evaluate an integral, we'll need a single beautiful theorem (The Fundamental Theorem of Calculus) plus various technical tricks by means of which we'll lure recalcitrant integrals into the Fundamental Theorem's sphere of influence.

* If, being older than twelve, you need a reminder, see exercise 1.

Example 1. Express the area of the shaded region in the figure as an integral.

Solution. We begin by thinking of the region as an infinite collection of infinitesimally thin rectangles. A typical rectangle (at a typical point x, as highlighted in the figure) has height $x^2/3$ and width dx, so its area is $(x^2/3)dx$.

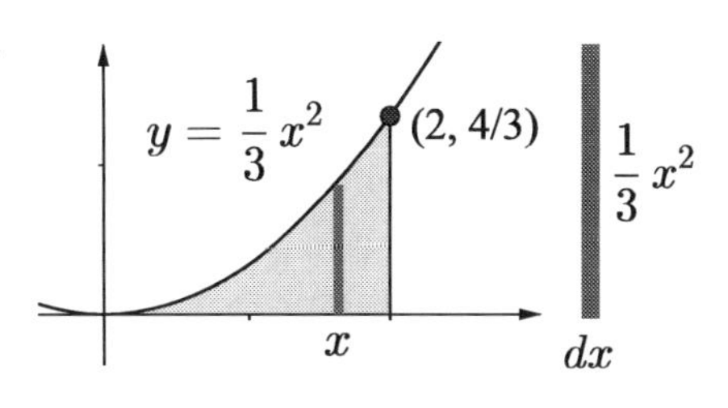

Hence, the integral

$$\int_0^2 \frac{x^2}{3}\,dx.$$

represents the shaded region's entire area. ♦

Once you've learned the Fundamental Theorem of Calculus, you'll be able to evaluate such integrals. But first, let's get comfortable interpreting and using integral notation. In the following examples, the goal is to understand *why* the integrals that eventually appear represent the quantities I claim they represent.

Example 2. Suppose that t hours after noon, a car's speed is $s(t)$ miles per hour. Express the distance that it travels between $2{:}00$ and $4{:}00$ pm as an integral.

Solution. For an object moving at constant speed, we know that $distance = speed \times time$.* Of course, during a typical two-hour drive, the car's speed is not constant; it can change wildly even within a single minute. Within a *second*, however, the car's speed can change only a bit. Within a tenth of a second, it can change still less, and within a hundredth of a second, its speed is nearly constant. As true calculus masters, we push this trend to extremes, recognizing that within any given *instant* (i.e. over an *infinitesimal* period of time), the car's speed actually *is* constant.

We'll now apply this crucial insight. At any time t, the car's speed is $s(t)$ *and remains so for the next instant*, so during that instant (whose duration we'll call dt), the car travels $\boldsymbol{s(t)dt}$ miles.

The distance that the car travels between $2{:}00$ and $4{:}00$ is the sum of the distances it travels during the (infinitely many) instants between those times – the sum, that is, of all of the $\boldsymbol{s(t)dt}$'s for all values of t between 2 and 4. Hence, the integral

$$\int_2^4 s(t)\,dt.$$

represents the total distance (in miles) that the car travels between $2{:}00$ and $4{:}00$ pm. ♦

Such is the spirit of integral calculus: We mentally shatter a function (whose values are *globally variable*) into infinitely many pieces, each of which is *locally constant* on an infinitesimal part of the function's domain. These constant bits are easy to analyze; having analyzed a typical one, we sum up the results with an integral, which the Fundamental Theorem will soon let us evaluate.

The overall process is thus one of *dis*integration (into infinitesimals) followed by *re*integration (summing up the infinitesimals) so as to reconstitute the original whole in a profoundly new form.

* E.g. A car driving at a constant speed of 60 mph for 2 hours would cover 120 miles.

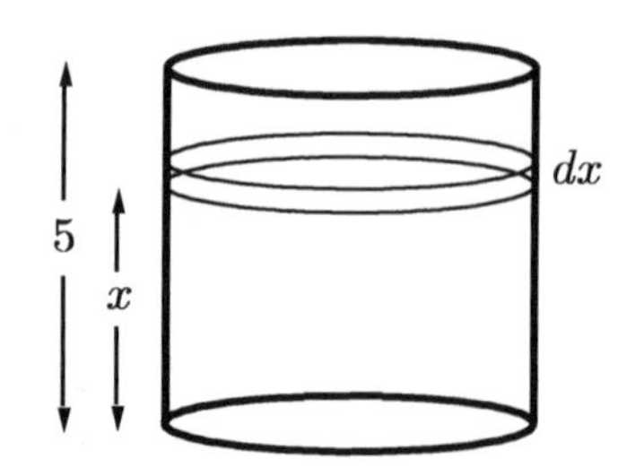

Example 3. In physics, the **work** needed to raise an object of weight W through a distance D (against the force of gravity) is defined to be WD. Suppose that a topless cylindrical tank, 5 feet high, with a base radius of 2 feet, is entirely full of calculus students' tears. Find an expression for the work required to pump all the salty liquid over the tank's top. (The density of the tears, by the by, is 64 pounds per cubic foot.)

Solution. Different parts of the liquid must be raised different distances. This *seems* to complicate our work (pun intended), but if we simply observe that all the liquid at the same height must be raised the same distance, an integral will save us. We simply don our calculus glasses and view the cylinder as a stack of infinitesimally thin cylindrical slabs (as though it were a roll of infinitely many infinitesimally thin coins). A typical slab at height x must be lifted through a *distance* of $5 - x$ feet. Since its *weight* is $256\pi dx$ lbs (64 times its volume of $\pi(2^2)dx = 4\pi dx$ cubic feet), the *work* needed to lift it to the tank's rim is $\mathbf{(5 - x)256\pi dx}$ foot-lbs.

To find the total amount of work required to remove all the slabs, we simply add up these $(5 - x)256\pi dx$'s as x runs from 0 (the tank's bottom) to 5 (the tank's top). Thus,

$$\int_0^5 (5 - x)256\pi \, dx.$$

is the total amount of work (in foot-lbs.) required to lift all the tears over the rim. ♦

Exercises.

1. Everyone knows that a rectangle with base b and height h has area bh.
Less well-known is that a *parallelogram* with base b and height h also has area bh.*

(Proof: In the figure at right, we cut a triangle away from a parallelogram and reattach it to the opposite side, turning the parallelogram into a *rectangle*. Since area was neither created nor destroyed during the operation, the original parallelogram's area must equal that of the rectangle, which is, of course, bh.)

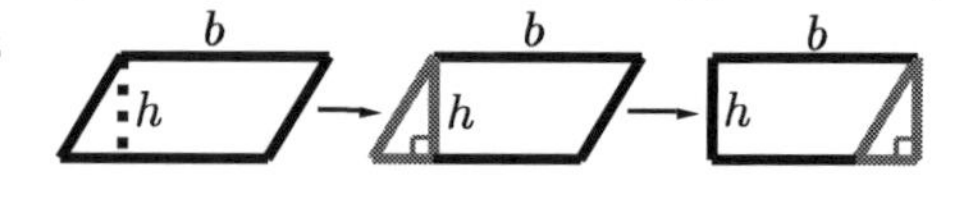

Your problem: Explain why **every triangle's area is *half* the product of its base and height**. [*Hint: We established the parallelogram's area formula by relating the parallelogram to a shape whose area we already knew. Use the same trick to establish the area formula for a triangle.*]

2. Express the following areas as integrals. In each case, sketch a typical infinitesimal rectangle.

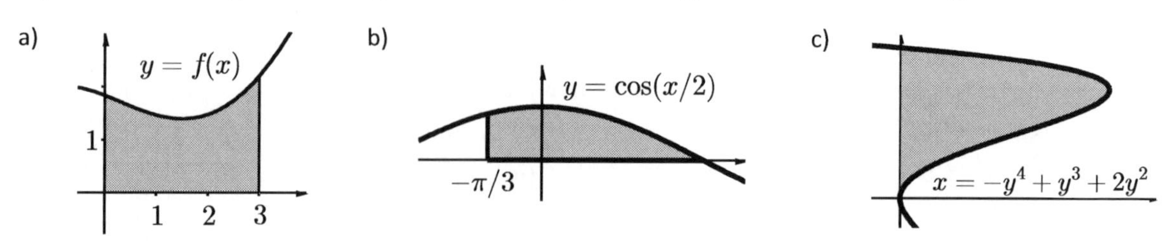

* A parallelogram's "height" is the distance between its parallel bases, *not* the length of its sloped sides.

3. Suppose the velocity of a particle moving along a straight line is, after t seconds, $v(t)$ meters per second.

a) Based on the graph at right, does the particle ever reverse its direction?

b) Can the total distance traveled by the particle between $t = 4$ and $t = 8$ seconds be expressed as an integral? If not, why not? If so, write down the integral that does the job, and explain *why* it represents this distance.

c) Observe that the integral you've produced also represents the area of the region that lies below the curve, above the x-axis, and between the vertical lines $t = 4$ and $t = 8$. The strange moral of this story: An area can sometimes represent a distance.

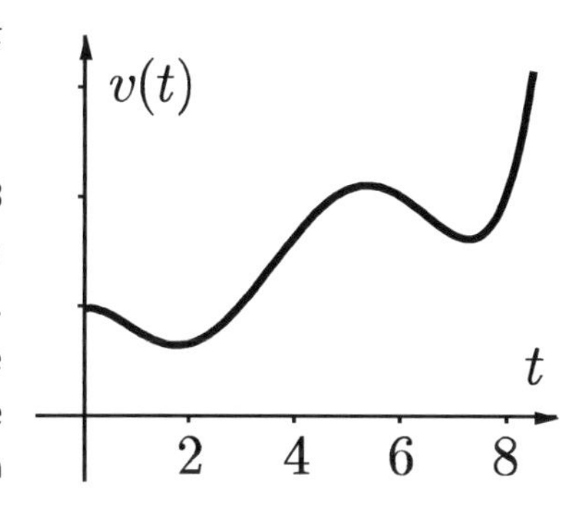

4. Draw the graph of $y = \sqrt{x}$ restricted to $0 \le x \le 3$. Look at your drawing and imagine revolving the graph around the x-axis. This will generate the three-dimensional "solid of revolution" shown at right. We can express its *volume* as an integral as follows. First, imagining the figure as solid, we mentally chop it up into infinitely many infinitesimally thin slices as though it were a loaf of bread. A schematic representation of one such slice is shown in the figure.

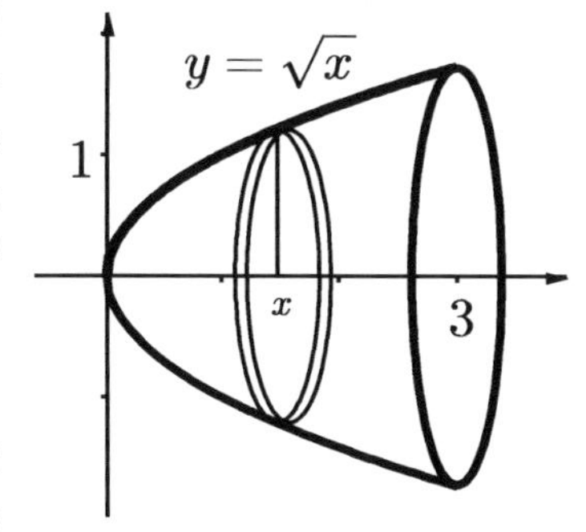

a) Explain why the solid's cross-sections will be circles.

b) The radii of these circles can vary from 0 to $\sqrt{3}$, depending on the point at which we take our slice. The closer the two circular cross-sections are to one another in space, the closer the lengths of their radii will be. When two circular cross-sections are *infinitesimally close* to one another, then their radii are effectively equal. Consequently, we can conceive of a typical infinitesimally thin slice as being an infinitesimally thin *cylinder*. This is excellent news because we know how to find a cylinder's volume: We multiply the area of its circular base by its height. In the specific example at hand, convince yourself that a typical slice of our solid (taken at a variable point x) is a cylinder whose circular base has radius $\sqrt{x}$ and whose height is dx (the slice's infinitesimal thickness). Then write down an expression for the volume of a typical slice at x.

c) Use your result from the previous part to express the volume of the entire solid as an integral.

d) If we revolve one arch of the sine wave around the x-axis, find an integral that represents the volume of the resulting solid. Draw pictures and let them guide you to the integral.

e) If we revolve half an arch of the sine wave (from 0 to $\pi/2$) around the $\boldsymbol{y}$-axis, draw a picture of the resulting solid, and write down an integral that represents its volume.

5. Express the shaded areas as integrals. [*Hint: The usual story. Disintegrate each area into infinitesimally thin rectangular slices, find an expression for the area of a typical slice, then integrate.*]

a)

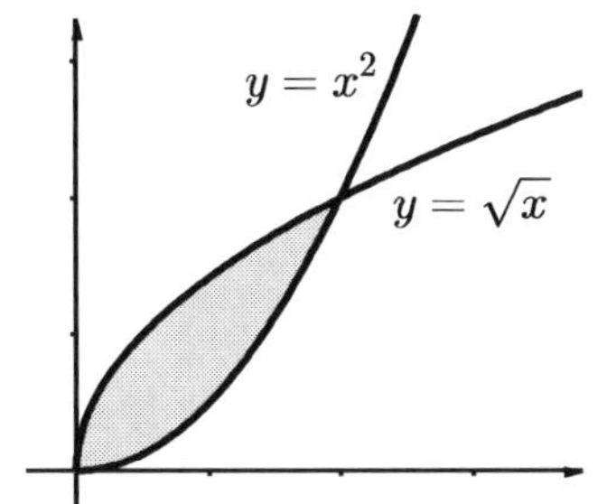

b)

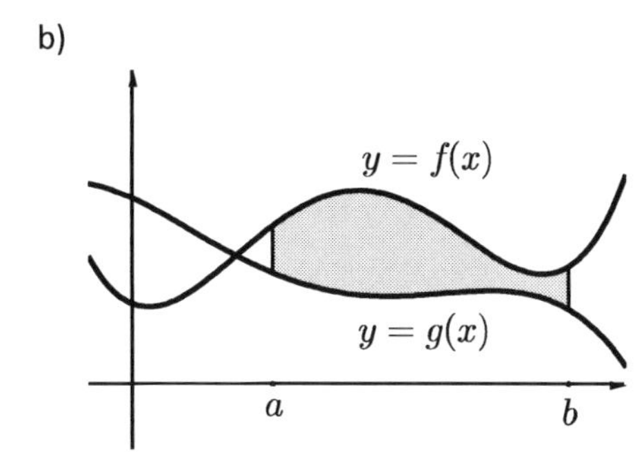

Positives and Negatives

"Natural selection can act only by the preservation and accumulation of infinitesimally small inherited modifications..."
- Charles Darwin, *On The Origin of Species*, Chapter IV.

Integrals always have the form $\int_a^b f(x)dx$. To interpret an integral, it sometimes helps to imagine an obscure mythological creature, the Integration Demon, who traverses the number line from a to b, pen and ledger in hand; at each real number in that interval, he evaluates f and multiplies the result by dx, yielding $f(x)dx$. His diabolical task is not just to find all the individual $f(x)dx$'s, but to *add them all up*. Their grand total is the integral's value.

When we integrate in the usual direction (left to right on the number line), dx is always a *positive* infinitesimal change in x.* Thus, for any x at which $f(x)$ happens to be *negative*, $f(x)dx$ is negative too. Naturally, a negative $f(x)dx$ goes, so to speak, into the debit column of the integration demon's ledger. If, at the end of the day, the debits outweigh the credits, then the integral's value will be negative.

Example 1. In the figure, it's easy to see that

$$\int_a^b f(x)dx = 6.$$

But what about the integral of f from b to c? Between b to c, each $f(x)dx$ will be *negative* since $f(x) < 0$ throughout that entire interval. Although the negative $f(x)dx$'s obviously can't represent areas (which are necessarily positive), they are still related to areas in a simple way.

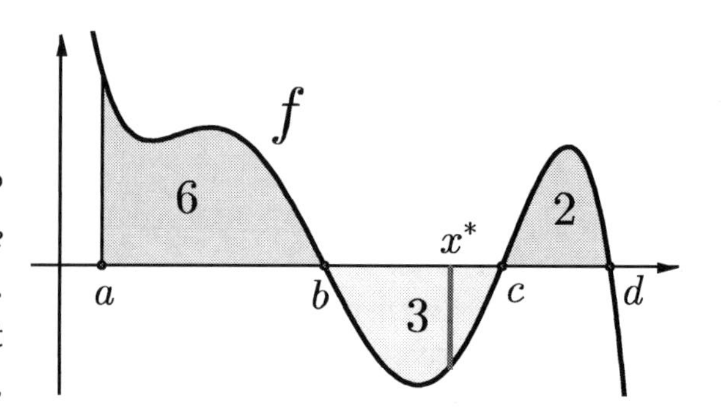

Consider the infinitesimal rectangle shown at x^*. Its height is $-f(x^*)$, so its area is $-f(x^*)dx$. This means that $f(x^*)dx$ must be the *negative* of that rectangle's area. Accordingly, as the integration demon goes from b to c, he'll be summing up not the areas of infinitesimal rectangles, but rather the *negatives* of their areas. Since these rectangles fill a region whose total area is 3, the demon's grand sum of all the $f(x)dx$'s between b to c must therefore be -3. That is,

$$\int_b^c f(x)dx = -3.$$

Next, suppose we wish to integrate from a to c. This integral's value is obvious if we think of the integration demon and his ledger. As he proceeds from a to b, he adds up lots of $f(x)dx$'s; their sum is 6 (the area of the figure's first shaded region). As he travels from b to c, he tallies up still more $f(x)dx$'s. Their sum, as discussed above, is -3, which brings his grand total down from 6 to 3. Consequently,

$$\int_a^c f(x)dx = 3. \qquad \blacklozenge$$

Please be certain that you understand the ideas that justify the preceding example's conclusions. If you do, then read on. If not, then go back, reread and think about them until you do understand.

* One *can* integrate "backwards" (*i.e.* right to left on the number line), but in practice, one doesn't. Should sheer perversity drive someone to do so, then his dx's would be *negative* infinitesimal changes. More on this in exercise 27.

Negative $f(x)dx$'s in an integral are perfectly acceptable in physical contexts, too, as we'll now see.

Example 2. A fly enters a room and buzzes around erratically, greatly irritating everyone therein. We'll concentrate exclusively on the fly's motion in the up/down dimension. Let $v(t)$ be the fly's upwards velocity after t seconds in the room. (Thus, $v(5) = -2$ m/s would signify that 5 seconds after entering the room, the fly's height is *decreasing* at a speed of 2 meters per second.)

Over a mere *instant* (i.e. infinitesimal bit of time), the fly's velocity is effectively constant. Hence, its *speed*, $|v(t)|$, is constant during the instant too, which means that during the instant, the constant-speed formula $speed \times time = distance$ applies. Thus, $|v(t)|dt$ represents the distance that the fly travels in the instant following time t. For example, the statement

$$\int_0^{60} |v(t)|dt = 10$$

tells us that during its first minute in the room, the fly travels a total of 10 vertical meters, sometimes going up, sometimes going down.

Whereas $|v(t)|dt$ is always positive and represents the distance that the fly travels in a given instant, the quantity $v(t)dt$ can be negative; it therefore indicates not only the distance travelled by the fly in a given instant, but the fly's *direction* during that instant as well. Positive $v(t)dt$'s correspond to altitude gains, while negative $v(t)dt$'s correspond to losses. It follows that when we sum up all the $v(t)dt$'s over some interval of time, we will end up with the fly's net gain in altitude over that time period, which could, of course, be negative. (Such a net change in an object's position is called its *displacement*.) Consequently, the equation

$$\int_0^{60} v(t)dt = -0.5$$

tells us that after sixty seconds of buzzing around (during which, according to the previous integral, it travelled a total of ten vertical meters) the fly's height was half a meter lower than it was when it first entered the room. ♦

After a few exercises, you'll be ready to begin climbing towards the Fundamental Theorem of Calculus, which will allow you to begin evaluating integrals at last.

Exercises.

6. In Example 1 above, explain why $\int_a^d f(x)dx = 5$.

7. The numbers in the figure at right represent areas. Evaluate the following integrals.

a) $\int_0^a f(x)dx$ b) $\int_a^b f(x)dx$ c) $\int_0^b f(x)dx$

d) $\int_a^d f(x)dx$ e) $\int_0^d f(x)dx$ f) $\left|\int_0^d f(x)dx\right|$

g) $\int_0^d |f(x)|dx$ [*Hint*: *Consider the graph of* $y = |f(x)|$.

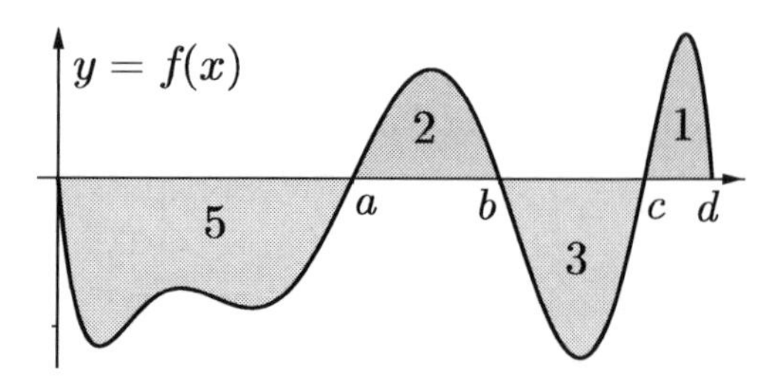

8. Suppose that after t seconds, a particle moving on a horizontal line has velocity $v(t)$ meters per second (where the positive direction is taken to be *right*). The graph of this velocity function is shown in the figure at right.

Explain the physical significance of each of the following three integrals:

a) $\int_0^2 v(t)dt$ b) $\int_0^5 v(t)dt$ c) $\int_0^5 |v(t)|dt$.

Now decide whether each of the following are true or false:

d) $\int_0^2 v(t)dt = \int_0^2 |v(t)|dt$. e) After five seconds, the particle is to the right of its initial position.

9. Find the numerical values of the following integrals by interpreting them in terms of areas. Draw pictures!

a) $\int_0^1 x\,dx$ b) $\int_{-1}^2 2x\,dx$ c) $\int_{-1}^1 \sqrt{1-x^2}\,dx$ d) $\int_0^2 -\sqrt{4-x^2}\,dx$ e) $\int_2^5 dx$ [Hint: $dx = 1dx$.]

10. a) Write expressions for the areas of each of the three infinitesimally thin rectangles depicted in the figure at right. Note: Even though all three expressions will turn out essentially the same, each case will require a slightly different justification. [*Hint: Bear in mind that while the values of functions may be negative, lengths are always positive*.]

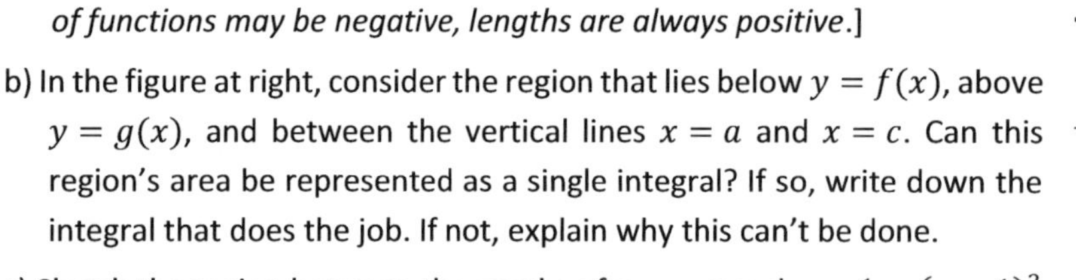

b) In the figure at right, consider the region that lies below $y = f(x)$, above $y = g(x)$, and between the vertical lines $x = a$ and $x = c$. Can this region's area be represented as a single integral? If so, write down the integral that does the job. If not, explain why this can't be done.

c) Sketch the region between the graphs of $y = -x$ and $y = 1 - (x-1)^2$ and express its area as an integral.

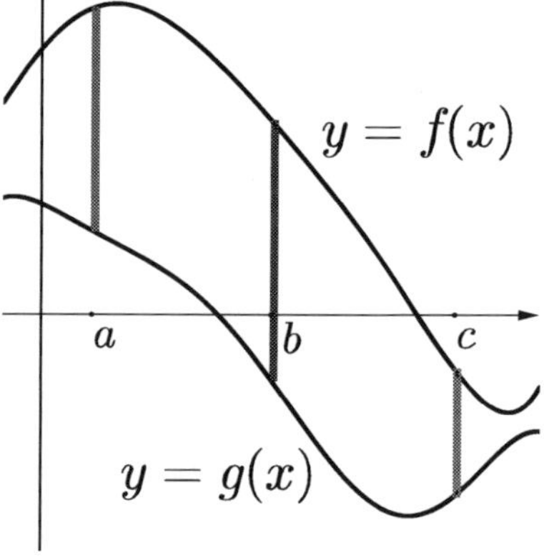

11. We say a function f is *even* if, for each x in its domain, $f(-x) = f(x)$, and *odd* if $f(-x) = -f(x)$ for each x.

a) Give some typical examples of even functions and odd functions, including trigonometric examples.

b) The algebraic definitions of even and odd functions have geometric consequences: The graphs of all even functions must exhibit a certain form of symmetry; the graphs of odd functions must exhibit another form. Explain the two forms of symmetry, and explain why they follow from the algebraic definitions.

c) Explain why the following is true: If f is even, then for any a in its domain, $\int_{-a}^a f(x)dx = 2\int_0^a f(x)dx$.

d) Explain why the following is true: If f is odd, then for any a in its domain, $\int_{-a}^a f(x)dx = 0$.

e) Is the function $f(x) = x^3 \sin^2 x - x\cos x$ even, odd, or neither?

f) Integrate: $\int_{-\pi}^{\pi} (x^3 \sin^2 x - x\cos x)\,dx$.

Preparation for the FTC: The Antiderivative Lemma

An **antiderivative** of a function f is a function whose derivative is f. (One antiderivative of $2x$, for example, is x^2. Another is $x^2 + 1$.)

Suppose we have two different antiderivatives of the same function. How different can they be? Well, by definition, their derivatives are equal, so their graphs change at *equal rates* throughout their common domain. When the graph of one increases rapidly, so does the graph of the other; when the graph of one decreases slowly, so does the graph of the other. This perfectly synchronized "dance of the antiderivatives" ensures that the distance between their graphs remains constant. (For it to change, the two graphs would have to change somewhere at *different* rates.)

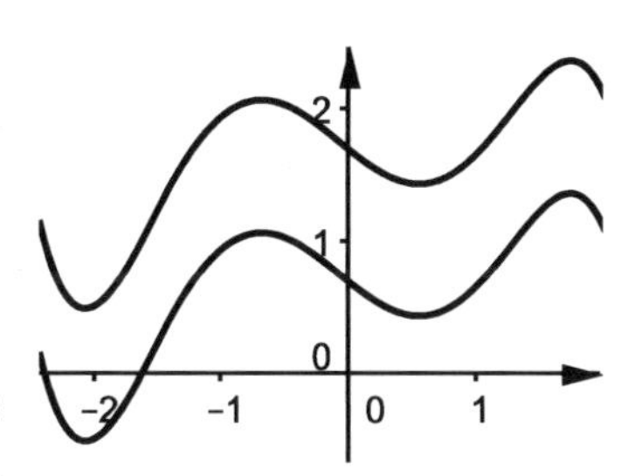

The moral of the story is the following lemma:*

Antiderivative Lemma. If g and h have the same domain and are antiderivatives of the same function, then they differ only by a constant. That is, $g(x) = h(x) + C$ for some constant C.

If we know one antiderivative of a function, we know them all: An antiderivative of $\cos x$ is $\sin x$; our lemma then guarantees that *every* antiderivative of cosine has the form $\sin x + C$ for some constant C.

Exercises.

12. An important rule you should immediately memorize: *To find an antiderivative of cx^r, where c is a constant, we increase its exponent by 1 and divide by the new exponent.*

a) Verify that an antiderivative of cx^r is $cx^{r+1}/(r+1)$, as claimed.

b) Write down antiderivatives for the following functions: x^3, $2x^9$, x, $-5\sqrt{x}$, $\sqrt[3]{x^2}$, $\frac{1}{x^2}$, $\frac{-8}{x^{3/7}}$, $x^{-\pi}$.

c) State the one value of r for which this rule does *not* apply. Why doesn't the rule apply in this case?

13. Think of an antiderivative of each of the following functions:

a) $\cos x$ b) $-\sin x$ c) $\sin x$ d) $\sec^2 x$ e) e^x f) $\sec x \tan x$ g) $\frac{1}{1+x^2}$ h) $-\csc^2 x$ i) $1/x$

14. You can sometimes find an antiderivative by guessing something close, then adjusting your guess to make its derivative come out right. For example, to find an antiderivative of $\sin(3x)$, we'd guess, "It will be something like $\cos(3x)$. But this function's derivative is $-\mathbf{3}\sin(3x)$, which is off by a constant factor of 3. To compensate, let's try multiplying our prospective *anti*derivative by $-1/3$. Will that work? Yes! A quick check shows that the derivative of $-(1/3)\cos(3x)$ is indeed $\sin(3x)$, so we've found our antiderivative."

Use this guess-and-adjust method to find antiderivatives of the following functions.

a) $\cos(5x)$ b) $\sin(\pi x)$ c) e^{-x} d) e^{3x} e) $\sec^2(x/2)$ f) 10^x g) 5^{-x} h) $\frac{1}{1+4x^2}$

15. Verify that $\ln(x)$ and $\ln(-x)$ are both antiderivatives of $1/x$, and yet they do *not* differ by a constant!

a) Why does this *not* violate our Antiderivative Lemma?

b) Is there an antiderivative of $1/x$ with the same domain as $1/x$? If so, what is it?

---

* A "lemma" is a small technical theorem used as a building block in the proof of a bigger, much more important theorem. After introducing one more preliminary idea, we'll use our lemma to establish the Fundamental Theorem of Calculus.

Preparation for the FTC: Accumulation Functions

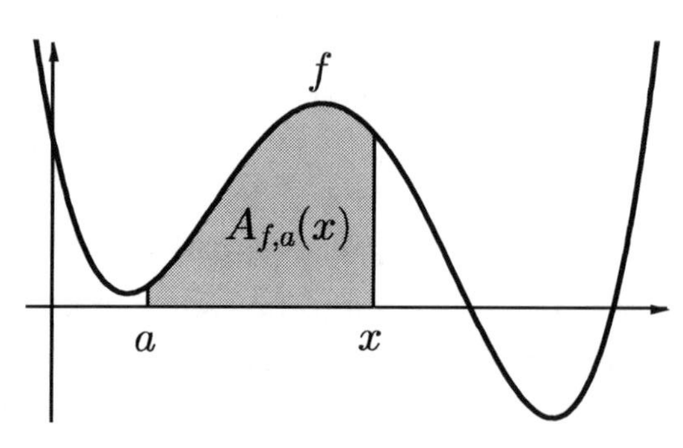

Take any function f and any fixed point a in its domain. From a, go down the horizontal axis to x, a variable point. In doing so, imagine "accumulating" all the area lying above the interval $[a, x]$ and below f's graph. The amount of area that we accumulate is clearly a function of x. We call this function, not surprisingly, the **accumulation function** $A_{f,a}$. (See the figure at right, which is worth a thousand words.)

If and when f's graph dips below the horizontal axis, the accumulation function *subtracts* any area it obtains from its running total, since any such area lies on the "wrong side" of the axis. Hence, an accumulation function's value will be positive or negative according to whether the majority of the area it accumulates lies above or below the horizontal axis.

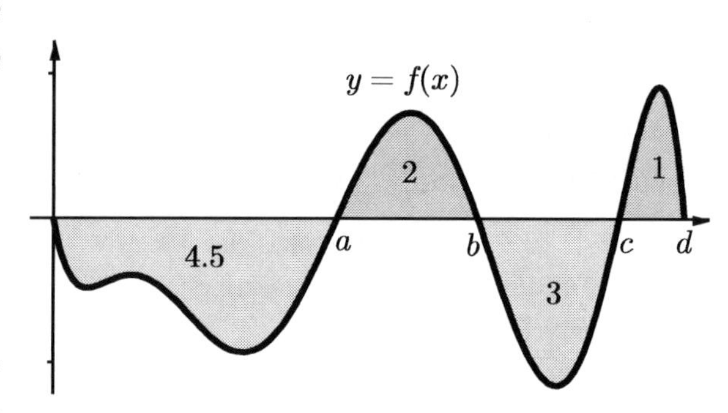

Example. Let the function f be defined by the graph in the figure at right. In this case, the following statements hold:

$$A_{f,a}(b) = 2, \quad A_{f,a}(c) = -1,$$
$$A_{f,a}(d) = 0, \quad A_{f,b}(c) = -3,$$
$$A_{f,b}(d) = -2, \quad A_{f,0}(b) = -2.5\,.$$

Once you've digested this example, you are ready to check your understanding further by doing the following exercises.

Exercises.

16. If $g(x) = 4 - x$, evaluate the following accumulation functions at the given inputs:

a) $A_{g,0}(4)$ b) $A_{g,2}(6)$ c) $A_{g,-2}(4)$ d) $A_{g,1}(1)$

17. If $s(x) = \sin x$, find the value of $A_{s,0}(2\pi)$.

18. We can express accumulation functions in integral notation. It is almost, but not quite, correct to write

$$A_{f,a}(x) = \int_a^x f(x)dx.$$

The problem is that this integral purports to sum things up while x runs from a to... x, which is syntactically incoherent. We stop summing when x is x? But when is x *not* x? To restore coherence, we recall that the variable in a function's *formula* is just a placeholder, a "dummy". (For example, it is just as reasonable to write the squaring function's formula as $f(t) = t^2$ as it is to write it as $f(x) = x^2$.) We therefore write

$$A_{f,a}(x) = \int_a^x f(t)dt,$$

which takes care of the problem, since we are now summing things up as $\boldsymbol{t}$ runs from a to x.

Your problem: If $g(x) = \cos x$, express the function $A_{g,0}(x)$ as an integral.

The Fundamental Theorem of Calculus (Stage 1: The Acorn)

"Only by considering infinitesimal units for observation (the differential of history, the individual tendencies of men) and acquiring the art of integrating them (finding the sum of these infinitesimals) can we hope to arrive at laws of history."

–Tolstoy, *War and Peace*, Epilogue, Part 2.

The Fundamental Theorem of Calculus develops in two stages, the first of which contains the second – in the sense that an unremarkable-looking acorn contains, *in potentia*, a mighty oak.

FTC (Acorn Version).
The derivative of any accumulation function of f is f itself.

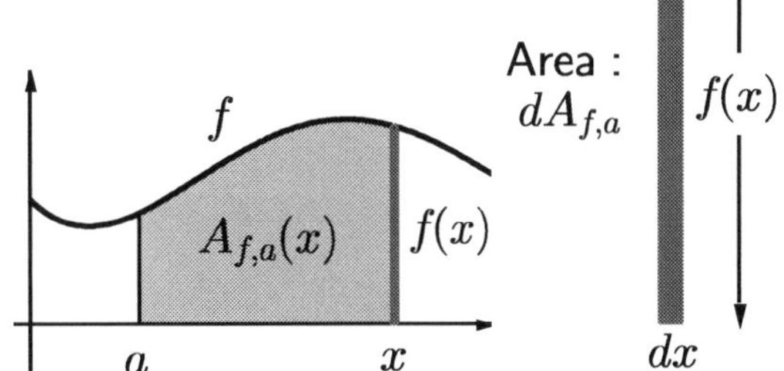

Proof. Let $A_{f,a}$ be an accumulation function for f. We'll find its derivative by the usual geometric procedure: We'll increase x by an infinitesimal amount dx, note the corresponding infinitesimal change $dA_{f,a}$ in the function's value, and finally, take the ratio of these two changes.

Since the accumulation function's value is the shaded area under the graph, its infinitesimal increase $dA_{f,a}$ (obtained by nudging x forward by dx) is the area of the infinitesimal rectangle I've emphasized in the figure. Dividing this rectangle's area by its width yields its height, $f(x)$. Rewriting this last sentence in symbols, we have

$$\frac{dA_{f,a}}{dx} = f(x).$$

This is exactly what we wanted to prove: The derivative of f's accumulation function is f itself. ■

The proof is simple enough, but having digested it, you probably remain puzzled by the result itself. After my grand promises that the FTC will help us evaluate integrals, I've presented you instead with a strange little acorn. What is it? What is it trying to tell us? Let us look closer.

Since an accumulation function is effectively an integral (a point made explicit in exercise 18), the acorn version of the FTC suggests that integration and differentiation are *inverse* processes. That is, if we start with a function f, take its integral (i.e. form an accumulation function from it), and then take the derivative of the result, we end up right back where we started – with f itself. Considering it this way, we recognize the acorn's latent power: It establishes a link between the well-mapped terrain of derivatives and the still-mysterious land of integrals. Just as we can use the simple properties of exponentiation to establish the basic laws of its inverse process (taking logarithms), we will now be able to use the simple properties of derivatives to crack the code of integrals.

To grow into an oak, our acorn still needs nutrients from the Antiderivative Lemma. In the following example, we'll bring the lemma and the acorn together at last. (As a fun but unimportant bonus, we'll also establish a surprising result about the sine wave.)

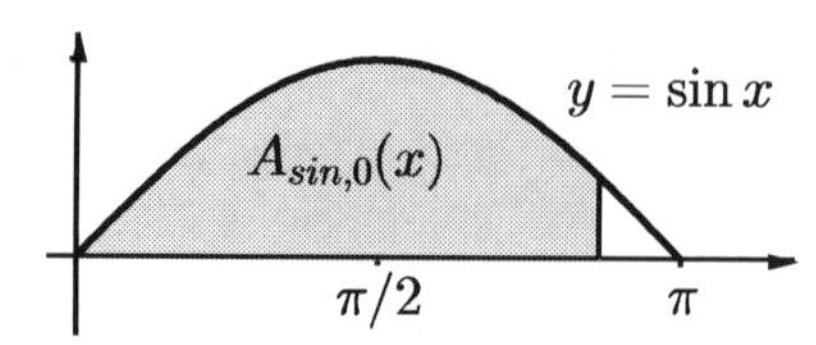

Example. Find the area under one arch of the sine wave.

Solution. The area we want is $A_{sin,0}(\pi)$. To find this value, we'll first produce an explicit formula for the accumulation function $A_{sin,0}(x)$, and then we'll substitute π into it.

To find our formula, we first note that $A_{sin,0}(x)$ is an antiderivative of $\sin x$ (by the FTC Acorn). Then, since we happen to know a formula for another antiderivative of $\sin x$ (namely, $-\cos x$), our Antiderivative Lemma guarantees that $A_{sin,0}(x) = -\cos x + C$ for some constant C. A-ha!

To find C's value, we substitute zero (our accumulation function's starting point) for x in the previous equation. Doing so shows that C must be 1, as you should verify. Hence, we now have

$$A_{sin,0}(x) = -\cos x + 1,$$

from which it follows that the area under one arch of the sine wave is exactly

$$A_{sin,0}(\pi) = -\cos \pi + 1 = 2! \quad \blacklozenge$$

That the area is a whole number is remarkable. Much more remarkable, however, is the argument by which we obtained this result. You'll use this argument in the exercises below. In the next section, we'll generalize it to produce the full FTC in all its oaky glory.

Exercises.

19. Sketch the region that lies below the graph of $y = (1/3)x^2$, above the x-axis, and between $x = 0$ and $x = 2$. Then, using the acorn version of the FTC (as in the example above), find its area.

20. Do the same for the region lying below $y = 1/(1 + x^2)$, above the x-axis, and between $x = 0$ and $x = \sqrt{3}$. [*Hint for the graph: Start with the graph of $y = 1 + x^2$. Think of how it would change if all of its points' y-coordinates were changed to their reciprocals; points far from the horizontal axis would be brought close to it, and vice-versa.*]

21. The acorn version of the FTC is often stated as follows: $\frac{d}{dx}\left(\int_a^x f(t)dt\right) = f(x)$.

a) Convince yourself that this symbolic statement is equivalent to the verbal statement of the Acorn FTC I've given above. (Recall exercise 18.)

b) Find the following derivatives: $\frac{d}{dx}\left(\int_a^x \sin(t)dt\right)$, $\frac{d}{dx}\left(\int_a^x \sin(t^2)dt\right)$, $\frac{d}{dx}\left(\int_a^{x^2} \sin(t)dt\right)$.

[*Hint for the third one: Let y be the function whose derivative you seek. Let $u = x^2$.* Then $\frac{dy}{dx} = \frac{dy}{du}\frac{du}{dx}$.]

22. Strictly speaking, our geometric proof of the Acorn FTC is not quite complete, though its holes are tiny and easily patched; you'll provide the necessary spackle in this problem.

a) The problem is that our proof tacitly assumed that $f(x) > 0$ at the point x where we took the accumulation function's derivative. Pinpoint the exact sentence in the proof in which we first used this assumption.

b) Having identified the flaw, we must show that the Acorn FTC holds even if we take the derivative at a point where $f(x) < 0$. To do so, draw a picture representing this situation. Since nudging x forward by dx will produce, in this case, an infinitesimal *decrease* in the accumulation function, we know that $dA_{f,a}$ will be the *negative* of the area of the rectangle in your picture. Translate this last statement into an equation; from it, deduce that even in this case, $dA_{f,a}/dx = f(x)$.

c) What if $f(x) = 0$ at the point where we take the accumulation function's derivative?

The Fundamental Theorem of Calculus (Stage 2: The Oak)

Once upon a time, you learned a clever technique for solving quadratics: completing the square. After using this technique a few times to solve particular quadratics, you applied it in the abstract to the equation $ax^2 + bx + c = 0$, thereby deriving the quadratic formula. With this formula in hand, you never again needed to complete the square to solve individual quadratics; the formula completes the square for you under the hood, automating the process so that you don't have to think about it.

Something similar is about to happen here. Having used the Acorn FTC to evaluate a few specific integrals, we'll now apply it to an integral in the abstract, thereby deriving the full-grown FTC, which we'll use thereafter to evaluate any integrals we meet, content to let it automate the details for us. We'll also drop the distinction between the two stages of the FTC; the oak contains the acorn as surely as the acorn the oak.

Recall the previous section's clever argument. We can think of an integral as representing one particular value of an accumulation function. By the Acorn FTC, this accumulation function is an *antiderivative* of the function being integrated. By the Antiderivative Lemma, we can express this accumulation function in terms of any other known antiderivative of the function being integrated. Consequently, *if we know another antiderivative* of the function being integrated, then we can find a formula for the accumulation function, which in turn allows us to evaluate the integral. It is a remarkable argument, and one that cries out for automation. Let us hearken unto its cries.

The FTC. If f is continuous over $[a, b]$ and has an antiderivative F, then

$$\int_a^b f(x)dx = F(b) - F(a).$$

Proof. Using the notation for accumulation functions introduced two sections ago, we note that the integral is equal to $A_{f,a}(b)$. To evaluate this expression (and thus to evaluate the integral), we'll find a formula for $A_{f,a}(x)$, and then let $x = b$.

By the Acorn FTC, we know that $A_{f,a}(x)$ is an antiderivative of f. Since F is an antiderivative as well, the Antiderivative Lemma assures us that $A_{f,a}(x) = F(x) + C$ for some constant C.

To find C's value, we let x be a in the preceding equation. Doing so yields $C = -F(a)$, so the equation in the preceding paragraph can be rewritten as $A_{f,a}(x) = F(x) - F(a)$. Consequently,

$$\int_a^b f(x)dx = A_{f,a}(b) = F(b) - F(a),$$

as claimed. ■

The Fundamental Theorem of Calculus tells us that when we integrate a function over an interval, the result depends entirely on the values of the function's antiderivative at the interval's two endpoints. This is – or at least it should be – astonishing: The interval consists of infinitely many points, and the integral is a sum of infinitely many terms... yet somehow the integral's value depends only on the antiderivative's value at *two* points? How can this be? I encourage you to sit by the fire some winter evening and meditate upon this mystery, bearing it in mind while thinking your way through the chain of proofs that led us to the Fundamental Theorem. The theme of understanding a function's behavior throughout a region by understanding a related function's behavior on the region's *boundary* will return when you study vector calculus; there, the regions you'll consider will be not just intervals of the one-dimensional real line, but regions of two, three, or higher-dimensional spaces.

From a pragmatic perspective, the FTC is of supreme importance because it lets us evaluate any integral once we know an antiderivative of the function being integrated. We simply evaluate the function's antiderivative at the boundaries of integration and subtract.

Example 1. Evaluate the integral $\int_2^3 x^4 dx$.

Solution. An antiderivative of x^4 is $F(x) = x^5/5$, so by the FTC,

$$\int_2^3 x^4 dx = F(3) - F(2) = \frac{243}{5} - \frac{32}{5} = \frac{211}{5}. \ \blacklozenge$$

To simplify our written work when evaluating integrals, some special notation has been developed:

$[F(x)]_a^b$ **is shorthand for** $F(b) - F(a)$.*

This handy bracket notation lets us dispense with explanatory phrases such as "Since $F(x)$ is an antiderivative of blah-blah-blah, the FTC tells us that…" when evaluating an integral. If, for instance, we redo the previous example with this notation, we can reduce the entire solution to just a few symbols:

Example 1 (Encore). Evaluate the integral $\int_2^3 x^4 dx$.

Solution. $\int_2^3 x^4 dx = \left[x^5/5\right]_2^3 = \frac{243}{5} - \frac{32}{5} = \frac{211}{5}$. ♦

This exemplifies the usual written pattern for using the FTC to evaluate simple integrals: We write down an antiderivative along with the attendant bracket notation, then we evaluate. Here's another quick example to ensure that the pattern is clear.

Example 2. Evaluate the integral $\int_{1/2}^1 \frac{2}{\sqrt{1-x^2}} dx$.

Solution. $\int_{1/2}^1 \frac{2}{\sqrt{1-x^2}} dx = \left[2\arcsin x\right]_{1/2}^1 = 2\arcsin 1 - 2\arcsin(1/2) = \pi - \frac{\pi}{3} = \frac{2\pi}{3}$. ♦

Polynomials will be especially easy to integrate once we've made two small observations. We've made the first already (in exercise 10): To find an antiderivative of cx^n, we raise the exponent by one and divide by the new exponent. The second is that *antiderivatives, like derivatives, can be taken term by term.*† Thus, to find an antiderivative of $x^3 + 2x^2 - 3x$, we simply sum up the terms' antiderivatives to obtain $(x^4/4) + (2x^3/3) - (3x^2/2)$. Armed with these observations, integrating a polynomial is trivial.

---

* The left bracket is often omitted by those for whom laziness trumps symmetry.

† Antiderivatives can be taken term by term *because* derivatives can. In symbols, if F and G are antiderivatives of f and g, then an antiderivative of $(f(x) + g(x))$ is $(F(x) + G(x))$ *because* $(F(x) + G(x))' = F'(x) + G'(x) = f(x) + g(x)$.

Example 3. Evaluate the integral $\int_0^2(-2x^2+4x+1)dx$.

Solution. $\int_0^2(-2x^2+4x+1)dx = \left[-\frac{2}{3}x^3+2x^2+x\right]_0^2 = \left(-\frac{16}{3}+8+2\right)-0=\frac{14}{3}$. ♦

Exercises.

23. Evaluate the following integrals, making use of the bracket notation introduced above.

a) $\int_0^{3\pi/2}\cos x\,dx$ b) $\int_1^2 -2x^3dx$ c) $\int_1^{e^2}\frac{1}{x}dx$ d) $\int_1^{\sqrt{3}}\frac{1}{1+x^2}dx$ e) $\int_{\pi/6}^{\pi/4}\sin x\,dx$ f) $\int_{-1}^{8}\sqrt[3]{x}\,dx$

g) $\int_0^{\pi/9}\cos(3x)\,dx$ [*Hint: Recall exercise* 14.] h) $\int_0^{\sqrt{3}/6}\frac{1}{\sqrt{1-9x^2}}dx$ i) $\int_0^1(x^9+5x^4-3x)dx$

j) $\int_0^{\pi/8}2\sin x\cos x\,dx$ [*Hint: Use a trigonometric identity to rewrite the function being integrated.*]

24. Find the areas of the shaded regions shown below:

a)

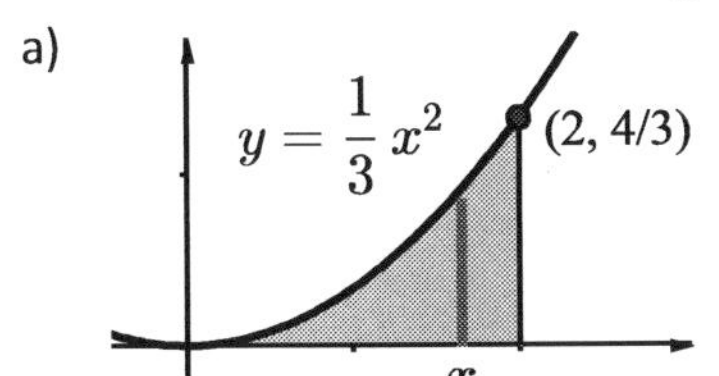

b)

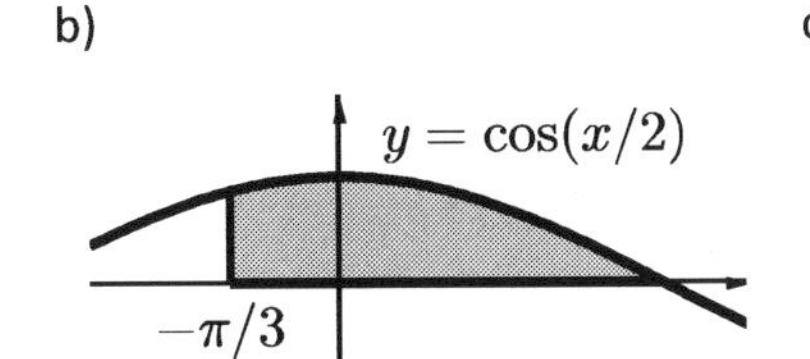

c)

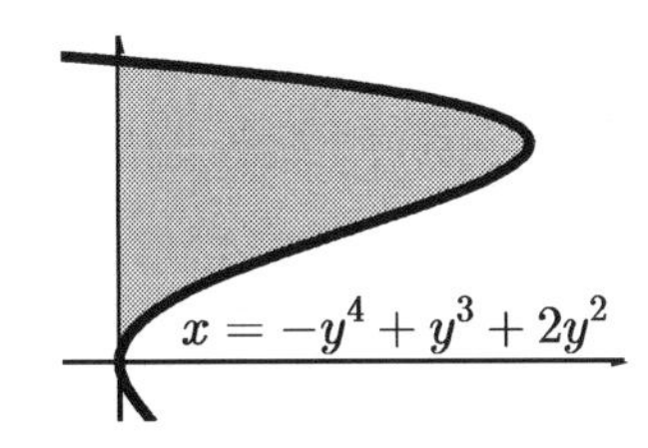

25. Consider the region that lies in the first quadrant, under $y=\sqrt[3]{x}$, and to the left of $x=8$. By revolving it about the x-axis (as in exercise 4), we obtain a solid of revolution. Sketch it, mentally decompose it into infinitesimally thin slices, then draw a typical slice and find its volume. Finally, find the volume of the full solid.

26. Consider the region lying in the first quadrant, above $y=\sqrt[3]{x}$, and below $y=2$. Revolve it around the $\boldsymbol{y}$-axis. Sketch it and so forth, ultimately finding its volume. [*Hint: To find the radius of a typical slice, you'll want to think of x as a function of y, rather than the other way around.*]

27. You probably know Democritus's famous theorem about cones: A cone takes up exactly $1/3$ of the space of the cylinder containing it. (See the figure.) Since a cylinder's volume is, of course, its base's area times its height, Democritus's result allows us to write down a formula for a cone's volume in terms of its base radius and height: $V_{cone}=(1/3)\pi r^2h$. Please do not clutter your memory with this formula; rather, just remember Democritus's result directly, from which you can reconstruct the formula in seconds whenever you need it.

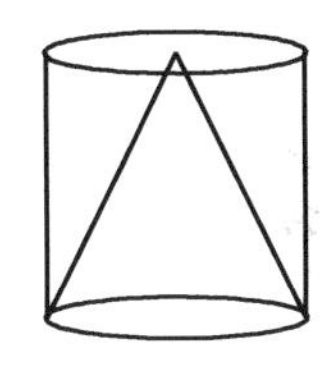

a) Prove Democritus right. [*Hint: Think of the cone as a solid of revolution generated by a line segment with one endpoint at the origin. Find the equation of the line containing the segment; naturally, it will involve r and h. Use calculus to find the generated cone's volume and verify that it agrees with the formula above.*]

b) Learn about Democritus, the laughing atomist philosopher.

28. Here is a sometimes useful fact: **Swapping an integral's boundaries of integration multiplies its value by -1.*** To see why this is so, return to first principles: To integrate is to sum up $f(x)dx$'s. If we integrate *backwards* (from right to left on the number line), we'll encounter the same values of $f(x)$ that we meet when integrating in the usual way, but the dx's – the infinitesimal changes in x – will be *negative*, since x *decreases* as we move leftwards on the number line. Thus, each $f(x)dx$ in the sum will be the negative of what it would have been had we integrated in the usual direction. The net result is that the value of our "backwards" integral is precisely the negative of its "forward" counterpart. *Your problem: Meditate upon this until it makes sense.*

---

* For example, since $\int_0^{\pi}\sin x\,dx=2$, our "useful fact" immediately tells us that $\int_{\pi}^{0}\sin x\,dx=-2$.

29. Rewrite the following as a single integral: $\int_a^b f(x)dx - \int_c^b f(x)dx + \int_c^d f(x)dx$. [*Hint: Exercise* 28 will help.]

30. Given the function f in exercise #7, evaluate the following integrals:

a) $\int_b^a f(x)dx$ b) $\int_c^b f(x)dx$ c) $\int_c^0 f(x)dx$.

31. a) Prove that a constant factor c can be pulled through the bracket notation. That is, $[cF(x)]_a^b = c[F(x)]_a^b$.

b) Use the fact that you can pull a constant multiple through the bracket while evaluating these integrals:

$\int_0^{3\pi/10} \cos(5x)dx$, $\int_0^1 e^{-5x}\,dx$, $\int_4^{25} \sqrt{x}\,dx$, $\int_0^{1/4} \frac{1}{1+16x^2}dx$

32. (The linearity properties of the integral)

a) Convince yourself that **constant factors can be pulled through integrals.**

$$\int_a^b cf(x)dx = c\int_a^b f(x)dx \text{ for any constant } c.$$

[*Hint: You can "see" this fact geometrically by thinking about rectangles' areas. Or think arithmetically: An integral is a sum; you should be able to convince yourself that the property above is essentially a matter of pulling out a common factor from each term in the sum.*]

b) Convince yourself that **the integral of a sum is the sum of the integrals.**

$$\int_a^b (f(x) + g(x))dx = \int_a^b f(x)dx + \int_a^b g(x)dx.$$

[*Hint: As in the previous part, you can "see" this fact geometrically by thinking about rectangles' areas. Or, think arithmetically: An integral is a sum; you should be able to convince yourself that the property above is essentially a matter of reordering the sum's infinitely many terms.**]

33. Is the integral of a *product* the product of the integrals? If so, prove it. If not, provide a counterexample.

34. A warning: *Vertical asymptotes cause the FTC to fail.* Should you try to integrate over an interval containing one, the result will be gibberish. For example, there is an infinite amount of area below the graph of $y = 1/x^2$ and between $x = -1$ and 1, but if we tried to apply the FTC to the integral $\int_{-1}^1 1/x^2\,dx$, we'd obtain this wholly erroneous "result":

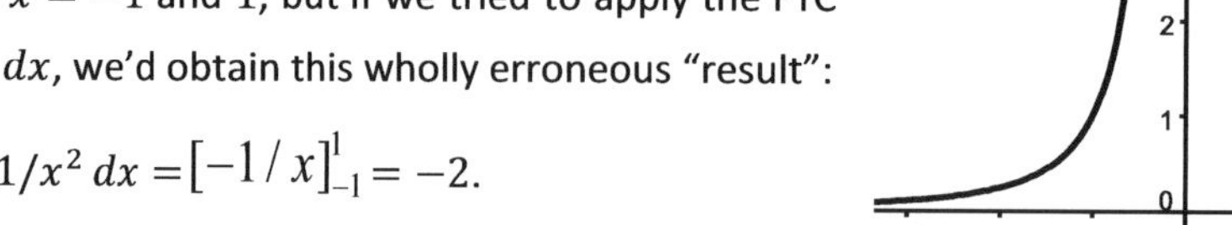

$$\int_{-1}^1 1/x^2\,dx = [-1/x]_{-1}^1 = -2.$$

Your problem: In the statement of the FTC in the text, a phrase warns you that the FTC does *not* apply to this integral. Identify the phrase.

* (A footnote you may safely ignore if you wish.) This innocent-looking statement may induce fits of violent swearing in tetchy calculus teachers. In Chapter 7 you'll learn why: Infinite sums don't always behave as innocently as their finite brethren; under some circumstances, reordering an infinite sum's terms can alter the sum's value. Fortunately, this disturbing phenomenon happens only in infinite sums of a sort never occurring in ordinary integrals, so it need not concern you here.

Still, if you are concerned (or are a tetchy calculus teacher yourself), you may prefer the following proof of the property in question: Let F and G be antiderivatives of f and g respectively. A linearity property of *derivatives* ensures that $F(x) + G(x)$ is an antiderivative of $f(x) + g(x)$. Hence, the FTC guarantees that

$$\int_a^b (f(x) + g(x))dx = [F(x) + G(x)]_a^b = (F(b) + G(b)) - (F(a) + G(a))$$

$$= (F(b) - F(a)) + (G(b) - G(a)) = [F(x)]_a^b + [G(x)]_a^b = \int_a^b f(x)dx + \int_a^b g(x)dx.$$

The preceding argument has the merit of demonstrating how to prove integral properties with the FTC by linking them to related derivative properties, but it conveys no real insight as to *why* the theorem holds. In contrast, "the theorem holds because it's just a rearrangement of the sum's terms" is both insightful and intuitive (after some initial thought), even if lacking in mathematical rigor. The goals at this stage should be insight and intuition; a healthy, astringent dose of rigor can always be added later. Rigor, injected too early, begets rigor mortis.

Chapter 5

The Integral Calculus Proper

Antiderivatives

"Do our dreams come from below and not from the skies? Are we angels, or dogs? Oh, Man, Man, Man! Thou art harder to solve than the Integral Calculus – yet plain as a primer."

- Babbalanja, from Herman Melville's *Mardi*, Chapter 136.

Heretofore, we've lacked a symbol for antiderivatives. To indicate that, say, cosine's antiderivative is sine, our sole recourse has been to write that very phrase. When "antiderivative" seemed an arcane concept, this lack of a symbol was acceptable. Those days are over. The FTC has revealed antiderivatives' centrality.

To indicate symbolically that one of $f(x)$'s antiderivatives is $F(x)$, we shall write $\int \boldsymbol{f(x)dx} = \boldsymbol{F(x)}$. Note well: *The antiderivative notation lacks boundaries of integration.*[*]

A function can't have just one antiderivative. If $f(x)$ has an antiderivative $F(x)$, then *all* functions of the form $F(x) + C$ (where C can be any constant) are antiderivatives, too. (Moreover, our "antiderivative lemma" guarantees there are no others.) To remind ourselves of this infinite family of antiderivatives, we traditionally include a "constant of integration" with our antiderivatives. For example, we usually write

$$\int \cos x \, dx = \sin x + \boldsymbol{C}.$$

In applied contexts, the constant of integration can carry vital information, as in the following example.

Example. From a tree branch 20 feet high, a squirrel flings a disappointingly unsavory chestnut to the ground with an initial downward velocity of 4 ft/second. How long is the nut in the air?

Solution. Let t be the number of seconds elapsed since the nut begins its fall. Thanks to Galileo, we know that the freely-falling nut's acceleration (in ft/s^2) must be the constant function

$$a(t) = -32.$$

Naturally, the nut's velocity function v is some antiderivative of its acceleration. That is,

$$v(t) = -32t + C$$

for some yet-to-be-determined constant C. To find C's value, we'll use the one known value of v ($v(0) = -4$); letting $t = 0$ in our velocity function yields $-4 = 0 + C$. Consequently, $C = -4$, so

$$v(t) = -32t - 4.$$ [†]

Next, we know that the nut's position (height) function is an antiderivative of its velocity. That is,

$$h(t) = -16t^2 - 4t + D$$

for some yet-to-be-determined constant D. To find D's value, we use our one known value of h (Namely, $h(0) = 20$). Letting $t = 0$ in the position function above yields $20 = D$, so it follows that

[*] That is, $\int f(x)dx$ represents a *function* (one of f's antiderivatives), whereas $\int_a^b f(x)dx$ represents a definite *number*, a sum. To distinguish them verbally, we call the former an "indefinite integral," and the latter a "definite integral."

[†] This argument exemplifies a common pattern in the subject of *differential equations*: An antiderivative reveals the *form* of some unknown function we seek (here, $v = -32t + C$), but it involves an unknown constant. To find the constant's value, we use an "initial condition" – a known value of our unknown function. Doing so reveals the function entirely. (Here, $v = -32t - 4$.)

$$h(t) = -16t^2 - 4t + 20.$$

Now that we've fully determined the nut's position function, a little algebra reveals that the nut hits the ground when $t = 1$. Hence, it spends exactly 1 second in the air. ♦

Because antiderivatives are called "indefinite integrals", methods for finding them are called "integration techniques". So far, we've employed just one: Stare at the function until an antiderivative occurs to you.* Clearly, we'll need more sophisticated techniques. In this chapter, you'll learn several. But first, exercises.

Exercises.

1. Find the indicated antiderivatives. Don't forget the constant of integration!

a) $\int (5x^4 - 8x^3)\,dx$ b) $\int \frac{1}{\sqrt{1-x^2}}dx$ c) $\int \cos(5x)\,dx$ d) $\int \frac{1}{x}dx$ e) $\int \frac{1}{5+x}dx$

f) $\int \frac{5}{1+z^2}dz$ g) $\int \frac{x^4+2x^3-x}{x^3}dx$ h) $\int \frac{x^2+\sqrt[3]{x}}{\sqrt{x}}dx$ i) $\int e^{-2t}\,dt$ j) $\int \frac{1}{3-2x}dx$

2. Each of the following statements is flawed. Identify the flaws.

a) $\int \cos x = \sin x + C$ b) $\int_0^1 2x dx = [x^2]_0^1 = (1^2 - 0^2) + C = 1 + C.$ c) $\int_0^1 \frac{1}{1+t^2}dt = \arctan t + C.$

3. Suppose we know that $f'(x) = 3x^2 - 3x + 1$ for some unknown function f.

a) Is this enough information to determine f completely? If so, write down a formula for f. If not, why not?

b) Suppose we also know that $f(1) = 2$. Now do we have enough information to determine f? If so, do so.

c) Explain *geometrically* why knowing the value of f at just one point sufficed to determine f completely.

4. Find f given that

a) $f'(x) = \sqrt{x}\left(1 + \frac{5x}{2}\right)$ and $f(4) = 32$.

b) $f''(x) = \sin x - \cos x$, $f'(0) = 3$, and $f(3\pi/2) = 1$.

c) $f''(x) = 4 + 6x$, $f(1) = 1$, and $f(-1) = 0$. [Hint: Find an expression for f involving *two* unknowns.]

5. From the roof of a 320 ft. building, Rollie Fingers hurls a baseball with an initial upwards velocity of 128 ft/sec.

a) Derive a formula for the ball's height (relative to the ground) t seconds after it leaves Fingers' fingers.

b) Find the maximum height the ball attains. c) How long does it take for the ball to reach the ground?

d) How fast is the ball moving when it hits the ground?

6. A rock falls from a cliff's edge. Shortly thereafter, it strikes the ground at 100 ft/s. How high is the cliff?
[*Hints: You know the rock's constant acceleration. Since it merely "falls", you know its initial velocity, too.*]

7. A particle appears on the x-axis at the number 5, and begins moving along it. Its initial velocity is 3.5 units/second, and after t seconds, its acceleration is $a(t) = \sin(2t)$. Find the particle's position function, $x(t)$.

8. Carson Carr's car careens down the road at 60 mi/hr. Suddenly, Mr. Carr stomps on the brakes as hard as he can, producing a constant deceleration of 20 ft/s^2. How far will his car travel before it stops?
[*Hints: Convert the car's speed to feet per second. Start the clock when he steps on the brakes.*]

9. (The linearity properties of indefinite integrals)

a) Explain why indefinite integrals can be taken term by term: $\int (f(x) + g(x))dx = \int f(x)\,dx + \int g(x)dx$.

b) Explain why a constant multiple can be pulled through an indefinite integral: $\int cf(x)\,dx = c\int f(x)\,dx$.

[*Hint: These are consequences of the linearity properties of* ***derivatives***.]

---

* This wonderful but severely limited technique includes an important variant, the "guess and adjust" game introduced in Chapter 4, exercise 14. (If you've forgotten this variant, you should review it; you'll need it in the exercises above.)

Substitution

Integrating a function by *substitution* entails changing its variable to simplify its form.* A skillful substitution can simplify an integral considerably.

Example 1. Use substitution to find $\int 10x\sqrt[3]{5x^2+7}\,dx$.

Solution. Step one in any substitution is to identify an expression in the integrand (the function we wish to integrate) whose derivative is in the integrand too. Call this expression u.

In the present example, we'll let $\boldsymbol{u = 5x^2 + 7}$ (since its derivative, $10x$, is in the integrand too).

Step two in any substitution: Using Leibniz notation, take u's derivative and then solve for du. Here, $du/dx = 10x$, so clearing the fraction yields $\boldsymbol{du = 10xdx}$.

The expressions we obtain for u and du constitute a "dictionary" with which we can translate the entire integral from the x-language to the u-language (and back again). Doing this here yields

$$\int 10x\sqrt[3]{5x^2+7}\,dx = \int \sqrt[3]{u}\,du = \frac{3}{4}u^{4/3} + C = \frac{3}{4}(5x^2+7)^{4/3} + C. \quad \blacklozenge$$

Such is substitution, which reveals, yet again, of the power of Leibniz's brilliant notation. To ensure that you understand the technique, use it to find $\int \tan x \sec^2 x\,dx$. Once you've succeeded (don't forget to check your answer by taking its derivative), you've grasped the basic idea.

Our guiding principle for choosing a substitution can be extended ever so slightly: *Let u be an expression in the integrand such that **a constant multiple of its derivative** also occurs in the integrand.*

Example 2. Find $\int xe^{3x^2}dx$.

Solution. If we let $u = 3x^2$, then its derivative, $6x$, doesn't occur in the integrand, but a constant multiple of it (namely, x), does. This is close enough. Behold:

If $\boldsymbol{u = 3x^2}$, then $du/dx = 6x$, so $du = 6xdx$. Hence, $\boldsymbol{xdx = (1/6)du}$. Translating everything into the "u-language", we find that

$$\int xe^{3x^2}dx = \int \frac{1}{6}e^u\,du = \frac{1}{6}e^u + C = \frac{1}{6}e^{3x^2} + C. \quad \blacklozenge$$

Finally, a word about substitution and *definite* integrals. In the expression $\int_a^b f(x)dx$, the x in $\boldsymbol{dx}$ reminds us that this is a sum as $\boldsymbol{x}$ runs from a to b. Consequently, if we translate the integral's variable from x to u (and thus translate from dx to du), **we must translate the boundaries of integration, too.** In other words, we must begin and end our sum at the $\boldsymbol{u}$-values that correspond to $x = a$ and $x = b$.

To translate the boundaries, we just ask, "What will u be when $x = a$? What will u be when $x = b$?" We obtain the answers from our dictionary, then switch the boundaries accordingly. Here's an example.

* You've seen such shenanigans in algebra. To solve $x^4 + x^2 - 1 = 0$, we let $u = x^2$, changing the equation to $u^2 + u - 1 = 0$. We find *this* equation's solutions and take their square roots (to convert u back to x) to get the original equation's solutions.

Example 3. Evaluate $\int_1^e \frac{\sqrt{\ln x}}{x}\,dx$.

Solution. Let $\boldsymbol{u = \ln x}$. Then $du/dx = 1/x$, so $\boldsymbol{du = (1/x)dx}$. These substitutions will simplify the integrand. To change the integration boundaries, we consult our boldface dictionary. It tells us that when $x = 1$, we have $u = \ln 1 = 0$. Similarly, when $x = e$, we have $u = \ln e = 1$. Thus,

$$\int_1^e \frac{\sqrt{\ln x}}{x}\,dx = \int_0^1 \sqrt{u}\,du = \int_0^1 u^{1/2}\,du = \left[\frac{2}{3}u^{3/2}\right]_0^1 = \frac{2}{3}. \quad \blacklozenge$$

When you can't think of an antiderivative by staring at the given function or by playing the "guess and adjust" game described in Chapter 4's exercise 14, then substitution *might* help. Always remember the guiding principle: Look for an expression in the integrand whose derivative (or a constant multiple thereof) is also in the integrand. Not all attempts at substitution work out well. Recognizing a promising candidate for "u" requires a practiced eye. Behold your chance to obtain that practice:

Exercises.

10. Find the following integrals.

a) $\int \sin x \cos x\,dx$ b) $\int x\sqrt{2x^2-1}\,dx$ c) $\int \frac{24x}{(6x^2+1)^2}\,dx$ d) $\int_0^{\sqrt{\pi}} x\cos(x^2)\,dx$ e) $\int_0^{\pi/4} \tan^2 x \sec^2 x\,dx$

f) $\int \frac{9x^2+2}{3x^3+2x+1}\,dx$ g) $\int \frac{\ln x}{x}\,dx$ h) $\int \frac{\arcsin x}{\sqrt{1-x^2}}\,dx$ i) $\int \frac{\cos x}{\sqrt{1+\sin x}}\,dx$ j) $\int_0^2 xe^{-x^2}\,dx$ k) $\int \frac{e^x}{1+e^{2x}}\,dx$

l) $\int_{\pi/10}^{\pi/5} \cos(5x)\,dx$ m) $\int_{-3}^{1} \frac{1}{\sqrt{3-2x}}\,dx$ n) $\int_0^3 \tan^2\left(\frac{x}{3}\right)\sec^2\left(\frac{x}{3}\right)dx$ o) $\int \frac{\sin(\sqrt{x})}{\sqrt{x}}\,dx$

11. Solving exercise 10a, David lets $u = \sin x$. He ultimately finds that the antiderivative he seeks is $(1/2)\sin^2 x$. Bathsheba, however, lets $u = \cos x$; she obtains the antiderivative $(-1/2)\cos^2 x$, as you should verify by doing the integral her way. David insists that Bathsheba must have done something wrong, for the answer in the back of the book is his, not hers, but she insists that her answer is correct. She tells David to take the derivative of her answer and see for himself. He does, and admits that *both* answers seem to be correct. "But doesn't that violate the Antiderivative Lemma?" he asks. Bathsheba, who understands these things well, replies, "Not at all." Explain.

12. Find $\int \tan x\,dx$. [*Hint: Use a well-known identity for tangent. Then use substitution*.]

13. Find $\int \cot x\,dx$.

14. a) Determine the point at which the graph at right crosses the x-axis.

b) If we revolve this graph around the x-axis, the result is a "solid of revolution" (as in Chapter 4, exercise 4). Find its volume.

c) The function graphed at right is undefined when $x = 0$. However, $\lim_{x\to 0^+} y$ *does* exist. Find its value.

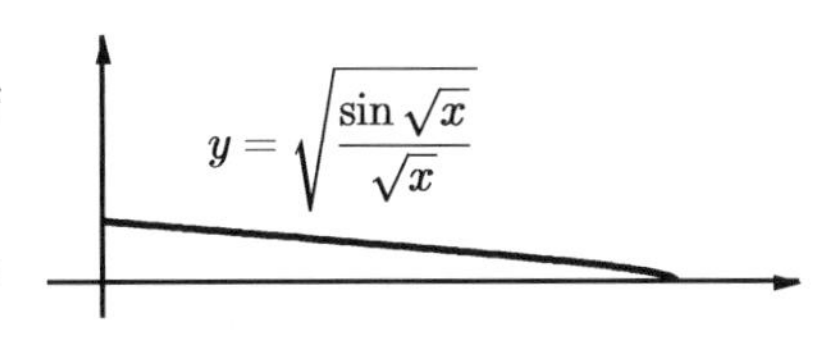

15. The antiderivatives of sine and cosine are obvious. Those of tangent and cotangent are easily obtained with substitution, as in exercises 12-13. Finding secant's antiderivative is much trickier. The clearest path begins:

$$\int \sec x\,dx = \int \frac{1}{\cos x}\,dx = \int \frac{\cos x}{\cos^2 x}\,dx = \int \frac{\cos x}{1-\sin^2 x}\,dx$$

a) Justify each step above. Note that the second move's strategic purpose was to set up that last integral, which a substitution will reduce to a simpler-looking integral devoid of trig functions. Make this substitution.

b) Try to integrate the function of u you've just obtained. You'll fail for now, but you'll soon learn how to do it.

Interlude with Elementary Functions

"Ask the local gentry,
and they will say it's elementary."
- Frank Sinatra, "Love and Marriage"

Elementary functions are functions built from simple parts in simple ways. Specifically, the "simple parts" are the basic precalculus functions (polynomial, exponential, logarithmic, and trigonometric), and the "simple ways" are finitely many function compositions and arithmetic operations ($\pm$, $\times$, $\div$, n^{th} roots). Thus, the realm of elementary functions comprises both simplicity, such as x^2, and complexity, such as

$$5e^x x^6 \ln x + \frac{\sin(2x)}{x^3} + \sqrt[3]{5\arctan(3x^2)}\,.$$

We rely on elementary functions without properly appreciating them. They are like the air we breathe. Rarely do we consider the possibility of *non*-elementary functions. (What would they even look like?)

While learning differential calculus, you remained safe and sound in the elementary functions' realm. The derivatives of polynomials, exponentials, logarithms, and trigonometric functions are all elementary. Moreover, the big structural theorems (linearity properties, plus the product, quotient, and chain rules) guarantee that all sums, differences, products, quotients, roots, and compositions of those basic functions have elementary derivatives, too. Hence, *all elementary functions have elementary derivatives.*

What about integral calculus? Do all the basic precalculus functions have elementary antiderivatives? Yes, they do.* This is promising, but it still doesn't quite guarantee that *all* elementary functions have elementary antiderivatives. For that, we'd need antiderivative analogues of the structural theorems for derivatives. Do such analogues exist? Alas, no. *There is no product rule for integration. There is no quotient rule for integration. There is no chain rule for integration.*†

Thus, when we try to integrate quotients, products, or compositions of functions, we are left to our own devices with no algorithm to guide us. Sometimes one technique will work, sometimes another will. We can, for example, integrate the particular quotient $(\ln x)/x$ with a substitution, but substitution won't work for *all* quotients. Integration is like a game: It involves experience, cleverness, sometimes even luck. The game's rules are simple: Given an elementary function, you must find an elementary antiderivative for it. But is there always a path to victory? In other words, does every elementary function have an elementary antiderivative?

Alas, no. *Many elementary functions do **not** have elementary antiderivatives.* To cite a few examples,

$$\frac{e^x}{x}, \qquad \frac{\sin x}{x}, \qquad \sin(x^2), \qquad \frac{1}{\ln x}, \quad \text{and} \quad e^{-x^2}$$

are innocent-looking elementary functions whose antiderivatives are *non*-elementary. To be sure, the FTC Acorn guarantees that these functions do have antiderivatives (namely, their accumulation functions), but mathematicians have proved that these antiderivatives *cannot* be expressed in elementary terms.

This does not mean that all accumulation functions are non-elementary. Some *are* elementary.

* At this stage, you can verify that polynomials, exponential functions, sine, cosine, tangent, and cotangent all have elementary antiderivatives. Soon, you'll be able to verify as much for secant and cosecant (using "integration by partial fractions"), and then for the logarithmic and inverse trig functions (using "integration by parts").

† There's no Santa Claus either, and life isn't fair.

Recall exercise 18 of Chapter 4, which shows how to express any accumulation function as an integral with a variable upper boundary. Sometimes, the FTC lets us convert an accumulation function from this integral form to an explicitly elementary form. For example,

$$A_{sin,0}(x) = \int_0^x \sin t\, dt = [-\cos t]_0^x = 1 - \cos x.$$

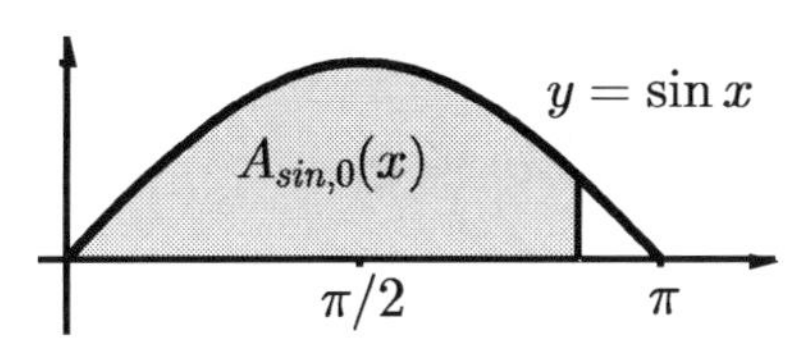

Since $A_{sin,0}(x)$ is just $1 - \cos x$ in disguise, we see that, despite its apparently non-elementary definition, it is elementary, after all.

And yet, let me repeat myself: *Many elementary functions do **not** have elementary antiderivatives.* The functions you'll integrate in this chapter's exercises have all been cherry-picked; they are functions that *do* have elementary antiderivatives. It is important that you develop the ability to express integrals in elementary terms when possible; it is equally important to recognize that this isn't always possible.*

Certain non-elementary antiderivatives of elementary functions are used in advanced mathematics. For example, although $1/\ln x$ has no elementary antiderivative, its *non*-elementary antiderivative $\int_0^x \frac{1}{\ln t} dt$ is massively important in the study of how the prime numbers are distributed among the whole numbers. Consequently, this function has been granted its own name (*the log-integral function*), and its own notation: It is denoted $\text{li}(x)$. Thus, the log-integral function is defined as follows:

$$\text{li}(x) = \int_0^x \frac{1}{\ln t} dt.$$

Just as you learned many properties of the sine function when studying trigonometry, so you will learn many properties of the log-integral function should you ever study analytic number theory. Just as you can consult a table, calculator, or computer to get approximate values of sine, so you can do the same to get approximate values of $\text{li}(x)$.†

Exercises.

16. The "hyperbolic cosine" is defined as follows: $\cosh x = \frac{e^x + e^{-x}}{2}$. Is it an elementary function? What is $\cosh 0$?

17. The "hyperbolic sine" is defined as follows: $\sinh x = \frac{e^x - e^{-x}}{2}$. What is $\sinh 0$? What is $\frac{d}{dx}(\sinh x)$?

18. Is the accumulation function $A_{x^2,2}(x)$ an elementary function? If so, express it in elementary terms.

19. To better appreciate my claim that elementary functions often lack elementary antiderivatives, try multiplying, dividing, and/or composing some basic precalculus functions, and then let *Wolfram Alpha* (or the computer program of your choice) try to integrate them. While playing around in this way, try to find some examples of elementary functions whose antiderivatives involve exotic functions such as the log-integral function discussed above, or whose antiderivatives cannot be expressed even in terms of these.

---

* Compare solving equations. Mechanical procedures leading to exact, closed-form solutions will take care of many equations, but not all. For example, $\cos x = x$ has a unique solution, but it can't be expressed in an exact, closed form. We can, however, approximate its value to any desired degree of accuracy. The same holds for definite integrals: If the function we wish to integrate has no elementary antiderivative, we can't produce the definite integral's exact value with the FTC, but we can still approximate its value to as many decimal places as we desire. More to come on this topic in Chapter 6.

† Typing $\text{li}(7)$ into the excellent online computer algebra system Wolfram Alpha, for example, we find that $\text{li}(7) \approx 4.757$.

Algebraic Trickery (Rational Functions)

"Fat-bottomed girls,
you make the rockin' world go round."
- Freddie Mercury

The sum, difference, or product of two polynomials is a polynomial, but the quotient of two polynomials is a so-called *rational function*. ("Rational" not because it is sensible, but because it is a ratio.) Polynomials are easy to integrate. Rational functions aren't. Rational functions do have elementary antiderivatives, but finding them can be awkward or even impossible in practice. Since a complete exposition on the subject of integrating rational functions is tedious and, frankly, of limited value, I'll confine myself to the simplest, most useful cases. A few preliminary paragraphs on rational functions will pave the way.

In deep ways, polynomials resemble integers, and rational functions resemble rational numbers.* In elementary school, you learned to divide integers and to note carefully the resulting remainder. You then learned that integer division lets us rewrite any rational number as an integer plus a *bottom-heavy rational* (e.g. $22/7 = 3 + 1/7$). In high school, you learned to divide polynomials and, once again, to note the remainder resulting from the process. You learned that polynomial division lets us rewrite any rational function as a polynomial plus a *bottom-heavy rational function*. (A "bottom-heavy" rational function is one whose denominator has a higher degree than its numerator.) For example, polynomial division reveals that

$$\frac{x^3 + 5x^2 + 3x + 5}{x^2 + 1} = (x + 5) + \frac{2x}{x^2 + 1},$$

as you should verify.†

This ability to rewrite any rational function as a polynomial plus a bottom-heavy rational is a boon for integration, for it lets us break the integral of any rational function into two simpler pieces: an integral of a polynomial, which is trivial, and an integral of a bottom-heavy rational. Thus, if we can integrate *bottom-heavy* rational functions, we can integrate all rational functions. Consequently, for the remainder of this section, we'll confine our attention to bottom-heavy rationals.

The simplest bottom-heavy rational is the reciprocal function $1/x$. The natural logarithm is its obvious antiderivative, but it is *incomplete*; the logarithm (unlike $1/x$) is undefined for negative values of x. **The complete antiderivative of $1/x$ is $\ln|x|$**: For positive x, it equals $\ln x$; for negative x, it equals $\ln(-x)$; for *both*, its derivative is $1/x$, thanks to the chain rule. Thus, like the reciprocal function, $\ln|x|$ is defined for all reals but zero, and its derivative is $1/x$ on its entire domain. That is,

$$\frac{d}{dx}(\ln|x|) = \frac{1}{x} \qquad \text{or equivalently,} \qquad \int \frac{1}{x}\,dx = \ln|x| + C.$$

Having recognized that the complete antiderivative of $1/x$ is $\ln|x|$, we can now integrate any bottom-heavy rational function *whose denominator is linear*. Clearly, any such function must have a constant for a numerator. To integrate any function of this form, we reason as follows:

---

* This resemblance is delineated in the area of mathematics known as *abstract algebra*.

† If you've forgotten (or never learned) how to do polynomial long division, please look it up online or in another book. The algorithm is almost exactly like the long division algorithm for whole numbers you learned on your mother's knee.

$$\int \frac{a}{bx+c}dx = a\int \frac{1}{bx+c}dx = a\left(\frac{1}{b}\ln|bx+c|\right) + C = \frac{a}{b}\ln|bx+c| + C.$$

Do not memorize this result. Just think it through until you can reproduce the line of thought leading to it. The only non-obvious step occurs at the second equals sign, and even that step is quite simple.

We're now ready to consider our main problem for this section: integrating a bottom-heavy rational whose denominator is *a product of* ***distinct*** *linear factors*. For example, suppose we wish to integrate

$$\frac{5x+2}{(x-2)(x+2)}.$$

This requires some algebraic trickery called "partial fraction decomposition". The idea is to rewrite the rational function above as a *sum* of two bottom-heavy rationals whose denominators are the original function's individual linear factors. Here, for instance, we wish to write

$$\frac{5x+2}{(x-2)(x+2)} = \frac{A}{x-2} + \frac{B}{x+2},$$

for some constants A and B.* If we can find such constants, we can integrate the function, for then each term of the right-hand side will be easy to integrate, as we saw in the previous paragraph.

How can we find the values of A and B? Sometimes through trial and error, but when this fails, then there's a more systematic way: Clear the fractions in the equation that contains them. Doing so here yields

$$5x+2 = A(x+2) + B(x-2).$$

This is an algebraic identity, so it holds for all values of x. Letting $x=2$, it yields $12=4A$, so $\boldsymbol{A=3}$. Letting $x=-2$ yields $-8=-4B$, from which it follows that $\boldsymbol{B=2}$. We've therefore discovered that

$$\frac{5x+2}{(x-2)(x+2)} = \frac{3}{x-2} + \frac{2}{x+2}.$$

As a check, we can add the terms on the right to confirm that their sum is the fraction on the left.

Note well: *The technique of partial fraction decomposition works* ***only*** *for bottom-heavy rationals.*†

Combining the ideas discussed throughout this section, we can now handle such integrals as this:

Example. Find $\int \frac{3x^2+5x-10}{x^2-4}dx$.

Solution.
$$\begin{aligned}\int \frac{3x^2+5x-10}{x^2-4}dx &= \int\left(3+\frac{5x+2}{x^2-4}\right)dx && \text{(polynomial division)}\\ &= 3x + \int\frac{5x+2}{(x-2)(x+2)}dx\\ &= 3x + \int\left(\frac{3}{x-2}+\frac{2}{x+2}\right)dx && \text{(partial fraction decomposition)}\\ &= 3x + 3\ln|x-2| + 2\ln|x+2| + C. \quad \blacklozenge\end{aligned}$$

---

* Had our original fraction's denominator been a product of 3 distinct linear factors, we'd have 3 terms on the right, with numerators A, B, C. The obvious generalization holds for 4, 5, or more distinct linear factors.

† Trying this technique on a rational function that *isn't* bottom-heavy can lead to disaster, as you'll see in exercise 24.

Exercises

20. Explain to someone else (or to an imaginary friend) why the complete antiderivative of $1/x$ is not $\ln x$, but $\ln|x|$.

21. Integrate:

a) $\int \frac{3}{x+5}dx$ b) $\int \frac{2}{5x-3}dx$ c) $\int_0^{\frac{e-1}{2}} \frac{3}{2x+1}dx$

22. Decompose into partial fractions:

a) $\frac{5x+7}{(x-1)(x+3)}$ b) $\frac{x}{x^2-5x+6}$ c) $\frac{3x^2+8x-4}{x^3-4x}$ [*Hint: Get distinct linear factors downstairs.*]

23. Rewrite each of these rational functions as the sum of a bottom-heavy rational and a polynomial. (If you've never learned polynomial long division, you can learn how to do it by looking it up online or in another book.)

a) $\frac{x+4}{x-1}$ b) $\frac{6x^2-2}{3x^2-x+1}$ c) $\frac{8x^4+2x^2+1}{2x^2+3}$

24. (For the algebraically curious.) I warned above that partial fraction decomposition works *only* for bottom-heavy rational functions. To see that this was no idle warning, let us try to apply partial fraction decomposition to a rational function whose top and bottom have the same degree. If we suppose that

$$\frac{x^2+1}{(x-1)x} = \frac{A}{x-1} + \frac{B}{x}$$

for some numbers A and B, we can "discover" that $A = 2$ and that $B = -1$, as you should verify by clearing the fractions and going through the usual motions. Unfortunately, adding the fractions $2/(x-1)$ and $-1/x$ reveals that their sum does *not* equal the original fraction, so this "discovery" is worthless. What happened?

What happened is that our argument proceeded from a false premise, rendering its conclusions worthless. We *supposed* our original fraction could be written in a specific form, and deduced that *if* it could be so written, then A and B would have to be 2 and -1 respectively. True enough... but it just so happens that our original fraction *cannot* be written in this form, making all subsequent deductions quite meaningless.*

Mathematicians have proved that bottom-heavy rational functions with distinct linear factors downstairs can *always* be decomposed into partial fractions in the manner described in this section. (The proof, however, is too gory for this family-friendly book.) Consequently, whenever these conditions are met, you can carry out partial fraction decomposition with nothing to fear.

Your problem: Vow to split *only* bottom-heavy rational functions into partial fractions!

25. Integrate:

a) $\int \frac{5x+7}{(x-1)(x+3)}dx$ b) $\int \frac{10-2x}{x^2+5x}dx$ c) $\int \frac{x^4}{x^2-4}dx$ [*Hint: Freddie Mercury.*]

26. If you ever study differential equations, you'll probably derive the *logistic function*, which is used to model an immense range of phenomena in fields as far-flung as ecology, economics, and medicine. The derivation involves the integral below (where m is a constant). Get a head start on your future; integrate it now.

$$\int \frac{1}{P(1-mP)}dP.$$

* To see *why* the original function can't be written in the specified form, suppose it could. Adding the fractions on the right would then yield a rational function with the desired denominator, $x(x-1)$, but whose numerator would be $(A+B)x - B$, as you should verify, which is hopelessly *linear* for any choices of A and B. Since no choices of A and B can yield the desired *quadratic* numerator, x^2+1, the specified form is one into which the given fraction cannot, alas, be put.

From a false premise, one can "deduce" literally anything. After making this point once in a public lecture, Bertrand Russell was challenged by a skeptical audience member to deduce from the false statement $0 = 1$ that he, Russell, was the Pope. Lord Russell's response: "If zero is one, then one is two. The Pope and I are two, so the Pope and I are one."

27. (The Antiderivative of Secant.) The antiderivatives of sine and cosine are obvious, while those of tangent and cotangent can be obtained in a matter of seconds by substitution, so there's no need to memorize them. In this exercise you'll find the antiderivative of secant.

a) First, the unsatisfying way: I'll tell you the answer. The antiderivative of secant is $\mathbf{\ln|\sec x + \tan x|}$. Verify this by showing that the preceding function's derivative is $\sec x$.

b) Second, we'll actually derive secant's antiderivative (as opposed to just taking its derivative.)* We'll begin as in exercise 15. Please redo that exercise, which shows that

$$\int \sec x \, dx = \int \frac{1}{1-u^2} du, \quad \text{where } u = \sin x.$$

c) Now that you know about partial fraction decomposition, use it to show that the integral above equals

$$\frac{1}{2}\ln\left|\frac{1+\sin x}{1-\sin x}\right| + C.$$

d) To put this in an equivalent form, multiply the top and bottom of the fraction within the absolute value by the conjugate of its denominator, $1 + \sin x$. Do this, and verify that the result can be written as

$$\frac{1}{2}\ln\left|\frac{1+\sin x}{\cos x}\right|^2 + C.$$

e) Finally, use a property of logarithms, and some very simple trig identities to rewrite this as

$$\ln|\sec x + \tan x| + C.$$

28. Derive the antiderivative of *co*secant by shadowing the argument we used in exercise 27b-e.

* Take note, sloppy users of terminology!

Trigonometric Trickery (Part 1)

Trigonometric identities can help us integrate many functions. Here are two well-known pairs of identities that often prove useful in calculus:

$\cos^2 x + \sin^2 x = 1$	and	$1 + \tan^2 x = \sec^2 x$	(Pythagorean identities)
$\cos^2 x - \sin^2 x = \cos(2x)$	and	$2 \sin x \cos x = \sin(2x)$	(Double-angle identities)

The two identities in the left column above look suspiciously similar. If we add (or subtract) them and then divide each side of the resulting equation by 2, we obtain the following half-angle identities:

$$\cos^2 x = \frac{1 + \cos(2x)}{2} \quad \text{and} \quad \sin^2 x = \frac{1 - \cos(2x)}{2} \qquad \text{(Half-angle identities)*}$$

These half-angle identities are probably new to you. You should commit them to memory, and – just as importantly – understand the simple method by which we derived them, so you can recover them should memory ever fail you. Here's an example of a half-angle identity in action:

Example 1. Integrate: $\int \cos^2(3x)\, dx$.

Solution.
$$\begin{aligned} \int \cos^2(3x)\, dx &= \int \frac{1 + \cos(6x)}{2} dx && \text{(by the half-angle identity for cosine)} \\ &= \frac{1}{2}\int (1 + \cos(6x)) dx \\ &= \frac{1}{2}\left(x + \frac{1}{6}\sin(6x)\right) + C \\ &= \frac{1}{2}x + \frac{1}{12}\sin(6x) + C. \end{aligned}$$
♦

Calculus teachers often put integrals of the form $\int \sin^m x \cos^n x\, dx$ on exams to see how well their students can juggle trig identities and substitution. If your teacher does this, you should rejoice, for these integrals are simple – once you get the hang of them. If either exponent is odd, then the trick that I'll use in the following example (or some minor variant thereof) will always work.

Example 2. Integrate: $\int \sin^4 x \cos^5 x\, dx$.

Solution.
$$\begin{aligned} \int \sin^4 x \cos^5 x\, dx &= \int \sin^4 x \cos^4 x \cos x\, dx && \text{(detaching the “odd” cosine)} \\ &= \int \sin^4 x\, (\cos^2 x)^2 \cos x\, dx \\ &= \int \sin^4 x\, (1 - \sin^2 x)^2 \cos x\, dx && \text{(Pythagorean identity)} \\ &= \int u^4 (1 - u^2)^2\, du, && \text{(Substitution: } u = \sin x) \\ &= \int (u^8 - 2u^6 + u^4)\, du \\ &= \frac{1}{9}\sin^9 x - \frac{2}{7}\sin^7 x + \frac{1}{5}\sin^5 x + C. \end{aligned}$$
♦

Be sure you understand the strategy behind that solution. After detaching an "odd" cosine factor, an even number of cosines remained, so we were able to convert them all to sines with the Pythagorean identity. This set the stage for an easy substitution. The same trick will work (with obvious changes) if we had begun with an odd number of *sine* factors: Detach one, then use the Pythagorean identity to convert the others to cosines. After that, you can finish the job with an easy substitution.

---

* Why the name? Let $x = \theta/2$ and then take square roots of both sides, and the reason will be clear. The form in which I've given the half-angle identities, however, is much more useful for integration.

Exercises.

29. Everyone knows the main Pythagorean identity, $\cos^2\theta + \sin^2\theta = 1$, but its alternate form can sometimes slip one's mind. Suppose you are taking an exam and can't recall it. How can you quickly derive it?

30. Explain how to derive the half-angle identities.

31. If we rotate the region shown at right about the x-axis, we produce a "solid of revolution" (as in exercises 4, and 25 - 27 in the previous chapter). This solid, however, is a bit different; when we cut a typical infinitesimally thin vertical slice of it, we obtain an infinitesimally thin circular cylinder *with a hole bored through it*. (Such a slice is sometimes called a *washer*, on account of its resemblance to the little ring of that name one finds on bolts and screws.)

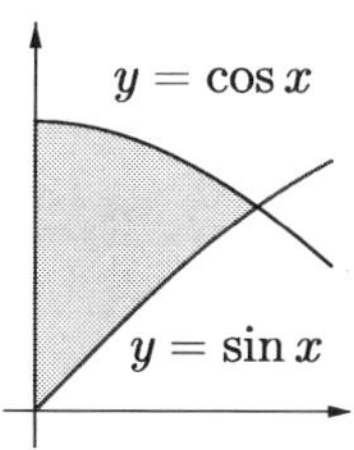

a) Sketch the solid of revolution, and a typical vertical slice thereof.

b) Despite its hole, a typical slice is the simplest sort of solid: all its cross-sections (in one direction) are identical. Consequently, its volume is just the area of any one of its cross-sections (equivalently, the area of its base) times its thickness. The area of its base is, as your figures should make clear, the difference of two circles' areas; its thickness is obviously dx. Find an expression for the volume of a typical slice.

c) Write down an integral expression for the volume of the solid of revolution.

d) Find the volume of the solid of revolution.

32. Integrate:

a) $\int \cos^2 x\, dx$ b) $\int_{\pi/8}^{\pi/4} 4\sin^2 x\, dx$ c) $\int \cos^2(2\theta)\, d\theta$

d) $\int_{\pi/12}^{\pi/4} \sin^2(4x)\, dx$ e) $\int (\cos^2 x - \sin^2 x) dx$ f) $\int_{\pi/4}^{\pi/3} (1 + \tan^2 x) dx$

g) $\int_0^{\pi} \sin x \cos x\, dx$ h) $\int \cos^2 x \sin^3 x\, dx$ i) $\int \sin^3 x \cos^6 x\, dx$

j) $\int \sin^3 x \cos^3 x\, dx$ [*Yes, **both** exponents are odd. Play around with it and you'll soon find a viable strategy.*]

k) $\int \sin^2 x \cos^2 x\, dx$ [*Half angle identities will help. A double angle identity could help, too.*]

l) $\int \tan^3 x \sec^6 x\, dx$

33. (A good workout for your trigonometry, analytic geometry, and calculus skills)

a) Sketch the portion of the graphs of $y = \tan x$ and $y = \cot x$ for which $0 < x < \pi/2$.

[*Hint: You know what the tangent graph looks like, of course. Cotangent is the reciprocal of tangent, so you should be able to use the known graph to sketch the unknown one. Just think about what would happen to the tangent graph if the y-coordinates of all its points were changed to their reciprocals.*]

b) Shade the shovel-shaped region that lies above the graphs you produced and below the line $y = \sqrt{3}$. Find the exact coordinates of the region's three vertices.

c) Revolve the region around the x-axis, and sketch the resulting solid.

d) Observe that a typical vertical slice of this solid will be a washer, as in exercise 31. Observe further that the expression for the volume of the washer at x could take two different forms: one when $\pi/6 \le x \le \pi/4$, another when $\pi/4 \le x \le \pi/3$. Write down expressions for each form.

e) To find the solid's volume, you could compute two separate integrals (based on the results of part (d)) and add their results, but a better way is to exploit the solid's obvious bilateral symmetry. Convince yourself that it suffices to compute only one of the two integrals; *doubling* its value will yield the solid's volume. With this in mind, write down an expression for the solid's volume that involves just one integral.

f) Find the solid's volume. [*Hint: To do the integral, the Pythagorean identity, in some form, will help.*]

Trigonometric Trickery (Part 2)

So far, we've used substitution to replace a relatively complex expression involving x with one symbol, u. In the technique of *trigonometric substitution*, however, we replace x with a far more complex expression: a trigonometric expression in θ. This may seem perverse, but under the right circumstances, taking one initial step backwards into complexity can help us take many subsequent steps towards our ultimate goal.

Trigonometric substitution, like completing the square, aims to rewrite part of an algebraic expression *as a square*. We usually apply trigonometric substitution to expressions lying under square roots.

Consider the expression $\sqrt{25-x^2}$. If we could somehow turn $25-x^2$ into a square, we'd be able to eliminate the radical. Anyone already experienced in the dark arts of trigonometric substitution knows how to do this: Let $x = 5\sin\theta$. For after we make this substitution, the expression becomes

$$\sqrt{25-25\sin^2\theta} = \sqrt{25(1-\sin^2\theta)} = 5\sqrt{1-\sin^2\theta} = 5\sqrt{\cos^2\theta} = 5\cos\theta,$$

and the square root is gone!* Very nice, but how would one ever think of that particular substitution?

Two ideas should be borne in mind when you are designing a trigonometric substitution. First, you are always trying to turn some expression or other (typically under a square root) into a square. Second, **the Pythagorean identities are our squaremakers**. To make $25-x^2$ into a square, we must think, "Can a Pythagorean identity help me turn 25 minus a square into a square?"

Yes. Rearranging the main Pythagorean identity yields $1-\sin^2\theta = \cos^2\theta$, which is halfway there: It turns $\mathbf{1}$ minus a square into a square. Multiplying both sides by 25 yields $25-25\sin^2\theta = 25\cos^2\theta$, which turns $\mathbf{25}$ minus a square into a square. A little meditation on this last equation makes the required substitution obvious: To turn $25-x^2$ into a square (namely, into $25\cos^2\theta$), we should let $x = 5\sin\theta$.

Once you've assimilated this way of thinking into your blood, you'll have little trouble with trigonometric substitution. Should you need to make a square from a *constant* ***plus*** *a square*, the alternate Pythagorean identity ($1+\tan^2\theta = \sec^2\theta$) will point you in the right direction. The same identity will also show you how to make a square out of a *square minus a constant*; just rearrange the identity into the form $\sec^2\theta - 1 = \tan^2\theta$, as in the following example.

---

* Algebraic fusspots may wonder, "Shouldn't there be an absolute value on that last cosine, since $\sqrt{x^2} = |x|$?" A fair criticism, but when I match your fussiness – and raise it – we will soon discover that the absolute value sign actually *isn't* necessary after all. I'll begin by upping the fussiness ante: Strictly speaking, the substitution that I made wasn't specific enough. I *should* have said, "Let $x = 5\sin\theta$, **where θ lies in the interval** $[-\pi/2, \pi/2]$." (We need the restriction to ensure a *unique* θ for any value of x.) Having clarified my earlier substitution, we can now observe that for every θ that lies in this restricted interval, $\cos\theta$ is positive. Consequently, for all such angles, we have $|\cos\theta| = \cos\theta$, so the absolute value sign turns out to be unnecessary, as claimed.

The preceding discussion, while certainly important, and perfectly suited to the snug privacy of the footnotes, would have been intolerably distracting in the body of the text. "The secret of being a bore," Voltaire noted, "is to tell everything."

Example. Integrate: $\int \frac{\sqrt{x^2-16}}{x}dx$.

Solution. Ordinary substitution won't help us here, but trigonometric substitution will. To eliminate the radical, we'll try to turn $x^2 - 16$ into a square. The Pythagorean identity $\sec^2\theta - 1 = \tan^2\theta$ turns a square minus 1 into a square. To find an identity that turns a square minus 16 into a square, we multiply both sides by 16, obtaining $16\sec^2\theta - 16 = 16\tan^2\theta$.

Our path is clear: To transform $x^2 - 16$ into a square (namely, into $16\tan^2\theta$), let $\boldsymbol{x = 4\sec\theta}$.

This substitution implies that $dx/d\theta = 4\sec\theta\tan\theta$, or equivalently, $\boldsymbol{dx = 4\sec\theta\tan\theta\, d\theta}$. Using the boldface equations to convert our integral from x to θ, we obtain, as you should verify,

$$\int \frac{\sqrt{x^2-16}}{x}dx = \int \frac{4\tan\theta}{4\sec\theta} 4\sec\theta\tan\theta\, d\theta = 4\int \tan^2\theta\, d\theta.$$

To integrate this last function, we can use the alternate Pythagorean identity again to write

$$4\int \tan^2\theta\, d\theta = 4\int (\sec^2\theta - 1)\, d\theta = 4\tan\theta - 4\theta + C.$$

Of course, we must now put this back in terms of the original variable, x. We'll do so by drawing a right triangle labelled to match our "translation equation", $x = 4\sec\theta$.* Looking at our triangle while intoning the name of SOH CAH TOA (the god of trigonometry), we see that

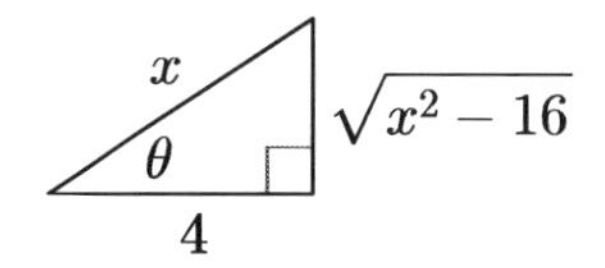

$$4\tan\theta - 4\theta = 4\left(\frac{\sqrt{x^2-16}}{4}\right) - 4\arctan\left(\frac{\sqrt{x^2-16}}{4}\right),$$

from which we conclude that

$$\int \frac{\sqrt{x^2-16}}{x}dx = \sqrt{x^2-16} - 4\arctan\left(\frac{\sqrt{x^2-16}}{4}\right) + C. \quad \blacklozenge$$

Such is the basic pattern for trigonometric substitution. Cumbersome, certainly, but remarkable: We've just used *trigonometry* to crack an integral that had, on its surface, no connection at all with that subject.

The flagpole and Ferris wheel problems of high school trigonometry leave one with little feel for how thoroughly the trigonometric functions pervade mathematics. As you climb Mt. Calculus, it is worth pausing now and again to look back at the plains of precalculus from whence you came. In doing so, you'll develop a greater appreciation for the elementary functions. This in turn may help you understand why we struggle to express antiderivatives in elementary terms when we can, even when the process is computationally laborious – as trigonometric substitution undoubtedly is.

* That is, we label one acute angle θ; then, since $\sec\theta = x/4$, we put the labels x and 4 on the hypotenuse and the leg adjacent to θ, respectively. By the Pythagorean Theorem, the remaining leg must be labelled $\sqrt{x^2-4}$. With the triangle thus completely labelled, we can read off that, say, $\tan\theta = \sqrt{x^2-4}/4$.

Exercises.

34. Suggest trigonometric substitutions for x that will turn the following expressions into squares. (Also, indicate the results that follow from making these substitutions.)

a) $9 - x^2$ b) $2 - x^2$ c) $x^2 + 16$ d) $x^2 + 15$ e) $x^2 - 3$ f) $9x^2 + 25$

35. Integrate, using the most efficient technique for the given integral (which may not be trig substitution):

a) $\int \frac{\sqrt{9 - x^2}}{x^2} dx$ b) $\int \frac{x^2}{\sqrt{9 - x^3}} dx$ c) $\int \frac{1}{\sqrt{4 + x^2}} dx$ d) $\int \frac{x}{\sqrt{x^2 - 4}} dx$ e) $\int \frac{1}{\sqrt{16 - x^2}} dx$

[*Hints: While working on (c), you'll eventually need secant's antiderivative, which we found in exercise 27. It's a good one to memorize. You can solve (e) with trig substitution, or perhaps by guessing and adjusting. If you try the latter approach, you might find it helpful to factor out a* 16 *under the radical before you guess.*]

36. While doing an integral (you have to know how to do these things when you are a king), David finds that the general antiderivative turns out to be $\ln|x/3| + C$. He turns to the answers in the back of the book, hoping to confirm his answer, but finds this instead: $\ln|x| + C$. Puzzled, he redoes the integral, but the pesky 3 remains part of his answer. He mentions it to Bathsheba, who tells him that both answers are correct, but the book's is preferable because it is cleaner. How can this be?

[*Hint: Recall the algebraic properties of logarithms. The logarithm of a quotient is…*]

37. Integrate: $\int \frac{1}{\sqrt{x^2 - 2}} dx$.

[*Hints: You'll need secant's antiderivative (from exercise 27). Simplify your answer using exercise 36's main idea.*]

38. Prove that the area of a circle of radius r is πr^2 by centering the circle at the origin, cutting it into infinitely many infinitesimally thin vertical strips, and summing up their areas.

39. Integrate: $\int \frac{1}{16 + x^2} dx$. [*Hint: See the hint for exercise 35e.*]

Integration by Parts

"I will heap mischiefs upon them; I will spend mine arrows upon them."
- Deuteronomy 32:23

The last integration technique we'll study bears the vague name of *integration by parts*, or IBP for short. We can apply it when the integrand is the product of two functions – one with a simple derivative, one with a simple antiderivative. It turns out that if we can integrate the product of the simple "parts" (the simple derivative and simple antiderivative), then IBP can help us integrate the original function.

Consider the integral $\int x \cos x\, dx$. Its integrand is the product of two functions, one of which, x, has a simple derivative, while the other, $\cos x$, has a relatively simple antiderivative. To organize our thoughts for IBP, we'll mark up our integral as follows:

$$\int \underset{\underset{1}{\downarrow}}{x} \overset{\overset{\sin x}{\downarrow}}{\cos x}\, dx .$$

The arrows point to derivatives, indicating that x's derivative is 1, and that $\cos x$'s antiderivative is $\sin x$. As explained above, if the product of the "parts" we've penciled in ($\sin x$ and 1) is a function that we can integrate (which is the case here), then IBP will help us integrate the original function.

I'll go ahead and show you *how* to integrate by parts before I explain *why* the technique works. The key is the **IBP Theorem: Any arrow-marked integral equals its "top product" minus its "outer integral".** "Top product" simply means the product of the functions at the *tops* of the arrows (here, x and $\sin x$), while the "outer integral" is the integral of the product of the parts we've penciled in *outside* the integral (here 1 and $\sin x$). Thus, according to the IBP Theorem, we have

$$\int \underset{\underset{1}{\downarrow}}{x} \overset{\overset{\sin x}{\downarrow}}{\cos x}\, dx = \underbrace{x \sin x}_{\text{"top product"}} - \underbrace{\int 1 \cdot \sin x\, dx}_{\text{"outer integral"}} = x \sin x + \cos x + C.$$

You may, of course, verify the truth of this statement by taking the derivative of the final expression.

Before we prove the IBP Theorem, here's another example, in which we'll use IBP *twice*.

Example. Integrate: $\int x^2 e^x dx$.

Solution. $\int \underset{\underset{2x}{\downarrow}}{x^2} \overset{\overset{e^x}{\downarrow}}{e^x}\, dx = x^2 e^x - \int \underset{\underset{2}{\downarrow}}{2x} \overset{\overset{e^x}{\downarrow}}{e^x}\, dx$ (IBP)

$= x^2 e^x - [2xe^x - \int 2e^x dx]$ (IBP again)

$= e^x(x^2 - 2x + 2) + C.$ ♦

Like every other integration technique, it takes practice to become comfortable with IBP. Once you've practiced a bit, though, you'll find it easy to use and – in contrast to partial fractions or trig substitution – refreshingly *clean*; you need only identify a promising pair of "parts" whose product is easier to integrate than your original integrand. Let the arrows guide you.

The IBP Theorem is a cousin of differential calculus' product rule. To understand their relationship, which will help us prove the IBP Theorem, we'll first need to restate the IBP Theorem in a symbolic form.

Our rhetorical statement of the IBP Theorem was: *Any arrow-marked integral equals its top product minus its outer integral.* Let us capture this in symbols. The integrand of an arrow-marked integral is a product of two functions – one with a simple derivative, one with a simple antiderivative. Calling the former function u and the latter function dv/dx (which makes v the simple antiderivative), we can write any arrow-marked integral as follows:

$$\int \underset{\substack{\downarrow \\ \frac{du}{dx}}}{u} \; \overset{\substack{v \\ \downarrow}}{\frac{dv}{dx}} dx .$$

Clearly, this makes the "top product" uv, and the "outer integral" $\int v\,(du/dx)dx$. Consequently, we may restate and prove the IBP theorem in the following symbolic form.

IBP Theorem (Formal Statement). If u and v are functions of x, then

$$\int u\frac{dv}{dx}dx = uv - \int v\frac{du}{dx}dx.$$

Proof. Since u and v are functions of x, the product rule tells us that

$$\frac{d}{dx}(uv) = u\frac{dv}{dx} + v\frac{du}{dx}.$$

Taking antiderivatives of both sides with respect to x, this becomes

$$uv = \int u\frac{dv}{dx}dx + \int v\frac{du}{dx}dx.$$

A little algebra shows that this is equivalent to

$$\int u\frac{dv}{dx}dx = uv - \int v\frac{du}{dx}dx,$$

which is precisely what we wanted to prove.* ■

Now that we've established the truth of the IBP theorem, you may confidently don your bow and quiver. You'll need them for the exercises that follow.

* If you cancel the dx's on both sides of the last equation, it becomes $\int udv = uv - \int vdu$. It is in this crude form that the IBP Theorem is usually stated by those benighted souls who know not the Braver Arrow Notation.

Exercises.

40. Integrate:

a) $\int xe^x dx$ b) $\int x^9 \ln x \, dx$ c) $\int x^2 \sin x \, dx$ d) $\int x \sin(3x+2) \, dx$ e) $\int \frac{\ln(\ln x)}{x} dx$

41. By now, you've learned how to integrate almost all the basic precalculus functions. The only ones you haven't integrated yet are the logarithmic and inverse trigonometric functions. We can finally dispose of this unfinished business with the help of IBP – plus an almost comically clever trick. The trick is simple: Insert a factor of 1. For example, to find $\int \ln x \, dx$, just rewrite the integral as $\int 1 \cdot \ln x \, dx$, then fire your arrows in the obvious way. Use this trick to integrate the following functions:

a) $\int \ln x \, dx$ b) $\int \arctan x \, dx$ c) $\int \arcsin x \, dx$ d) $\int \arccos x \, dx$

42. (A self-replicating integral) If we apply IBP to $\int e^x \cos x \, dx$, the resulting "outer integral" will be $\int e^x \sin x \, dx$, which is no easier than the original integral. Interestingly, if we apply IBP to the outer integral, then *its* outer integral is the *original* integral, $\int e^x \cos x \, dx$. We are going around in circles, yes, but from this circular motion we can harness enough energy to integrate the original function. Here's how:

Express the two applications of IBP described above as a mathematical sentence, involving two equals signs. Your sentence should let you conclude that $\int e^x \cos x \, dx = (\text{some stuff}) - \int e^x \cos x \, dx$, which you can think of as an algebraic equation *whose unknown is the integral.* Now use basic algebra to solve for the integral!

43. (Clever, very.) Despite all the techniques you've learned, if you try to integrate $\int (\arcsin x)^2 dx$, you probably won't get anywhere. Take a few minutes and try. Perhaps the integrand has no elementary antiderivative? Oh, it has one, all right, but the path to it well-hidden.

A sly substitution will work: Let $x = \sin\theta$. This reduces the integral to something you can handle with IBP. Carry out the details to find $\int (\arcsin x)^2 dx$.

44. Revolve the first arch of the sine wave (from 0 to π) around the $\boldsymbol{y}$-axis. Find the volume of the resulting solid by cutting it into *washers* as in exercise 31.

45. You've found volumes of solids of revolution in earlier problems by slicing them into cylindrical slabs or washers. Another method is to think of such a solid as being composed of infinitely many infinitesimally thin *cylindrical shells*, as in the image at right.* If we could express the volume of a typical shell (crossing the positive x-axis at point x), then we could integrate as x runs from 0 to b.

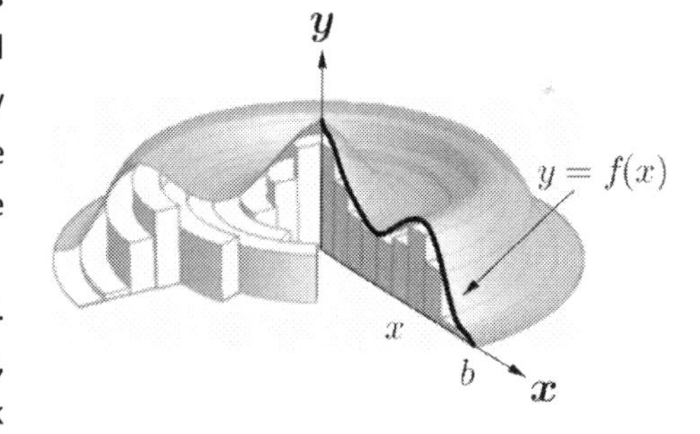

Mentally isolate the shell that crosses the positive x-axis at x. To find its volume, slice it parallel to the y-axis, then "unroll" it, flattening it out into an infinitesimally thin rectangular slab. (Think of cutting a paper-towel roll and spreading it flat on a table.)

a) Draw a typical shell and the slab it becomes after surgery.

b) Explain why the slab's dimensions are $f(x)$, $2\pi x$, and dx.

c) Use the ideas in this problem to find the volume of the solid in exercise 44. You'll find this method easier.

d) Revolve the region bounded by $y = \sqrt{x}$ and $y = x^2/8$ about the y-axis. Find the resulting solid's volume.

e) Use cylindrical shells to establish the formula for the volume of a sphere of radius r.

f) Use cylindrical shells to establish the formula for the volume of a cone of height h and base radius r.

---

* The original image was created by Blacklemon67 at English Wikipedia; I alone am responsible for defacing it with axes and labels. The surface and shells, of course, are supposed to wrap entirely around the y-axis.

Chapter 6

Odds and Ends

Arclength

"Have mercy upon me, O Lord; for I am weak:
O Lord, heal me; for my bones are sore vexed.
My soul is also sore vexed: But thou, O Lord, how long?"
- Psalms 6:2-3

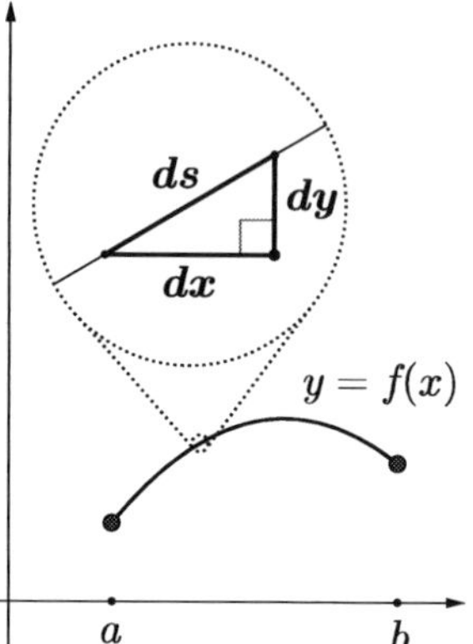

How can we find the length of a curve? If you have imbibed the spirit - and not merely the algorithms - of integral calculus, the strategy will not surprise you: We imagine the curve as a chain of infinitely many infinitesimal line segments. We then find an expression for a typical segment's length. Finally, we sum up these infinitesimal bits of length with an integral.

If s represents arclength, we'll need an expression for ds, an infinitesimal bit of arclength. The figure at right ("the differential triangle") suggests that

$$\boldsymbol{ds^2 = dx^2 + dy^2},$$

but since squared infinitesimals are zeros, this tells us nothing. However, if we divide each of the triangle's sides by dx, we obtain a similar triangle with *real* side lengths: 1, dy/dx, ds/dx. Applying the Pythagorean Theorem to *that* triangle and solving for ds yields, as you should verify:

$$ds = \sqrt{1 + \left(\frac{dy}{dx}\right)^2}\, dx.$$

This expression for ds allows us to find arclengths by integrating with respect to x (as x runs from a to b). An example will make this clear.

Example. Find the length of the graph of $y = x^{3/2}$ between $(0,0)$ and $(4,8)$.
Solution. By the formula above, our infinitesimal element of arclength ds is

$$ds = \sqrt{1 + \left(\frac{3}{2}\sqrt{x}\right)^2}\, dx = \frac{1}{2}\sqrt{4 + 9x}\, dx.$$

Consequently, the total arclength is given by

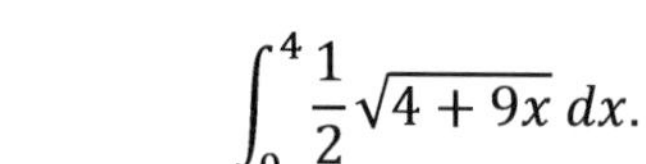

$$\int_0^4 \frac{1}{2}\sqrt{4 + 9x}\, dx.$$

As you should verify, this integral yields a total arclength of $8\left(10\sqrt{10} - 1\right)/27$ units. ♦

Exercises.

1. Find the lengths of the following curves over the specified intervals.
a) $y = (1/3)(x^2 + 2)^{3/2}$, $0 \le x \le 2$. b) $y = \ln(\sec x)$, $0 \le x \le \pi/4$. [*Hint: Ch. 5, Exercise 27.*]

2. a) Explain to a friend *why* the formula given above for ds holds.
b) Dividing the differential triangle's sides by $\boldsymbol{dy}$ yields another scaled-up real version. Draw it, labelling its sides.
c) Use your triangle from (b) to explain why $ds = \sqrt{(dx/dy)^2 + 1}\, dy$.
d) Use the formula in part (c) to find the length of $x = (2/3)(y - 1)^{3/2}$ over the interval $1 \le y \le 4$.

Surface Area

If we slice off a cone's top, cutting parallel to its base, the resulting solid is called a conical *frustum*. It is clearly a solid of revolution. The following geometric formula, which you'll prove in the exercises, gives a frustum's lateral surface area in terms of its *generator* – the revolving line segment that traces it out:

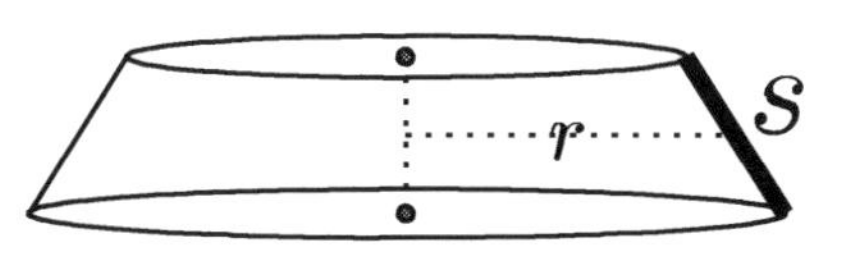

$$\begin{pmatrix}\text{Lateral surface area}\\ \text{of a conical frustum}\end{pmatrix} = \begin{pmatrix}\text{Length of its}\\ \text{generator}\end{pmatrix}\begin{pmatrix}\text{Distance travelled by its}\\ \text{generator's midpoint}\end{pmatrix}.$$

Thus, the frustum in the figure above would have a lateral surface area of $s(2\pi r)$.

Conical frusta arise when we compute a surface of revolution's *surface area*. If we think of a curve as a chain of infinitesimal line segments, then when we revolve it around an axis, each infinitesimal "link" in the chain will trace out an infinitesimally-thin conical frustum. Thus, if we can find an expression for the surface area of a typical frustum, we can obtain the area of the full surface by integrating.

Example. Revolving $y = \sqrt{x}$ (for $0 \leq x \leq 4$) about the x-axis yields a surface. Find its surface area.

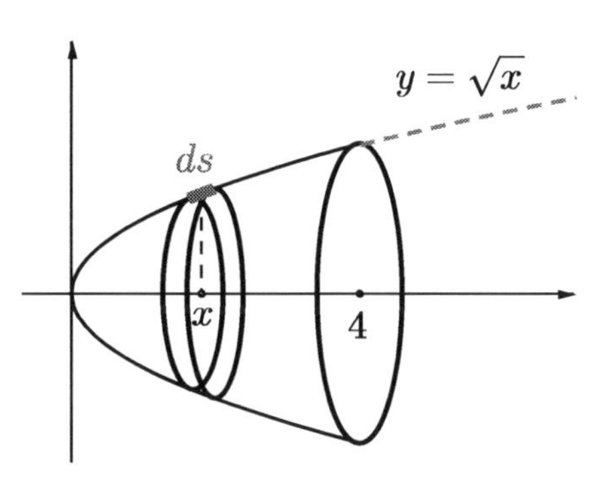

Solution. Think of the surface as consisting of infinitely many infinitesimally thin conical frusta. A schematic drawing of a typical frustum at x is highlighted in the figure at right.

According to the discussion above, our typical frustum's lateral surface area is $(2\pi\sqrt{x})ds$, since it is generated by a segment of length ds, whose midpoint travels along a circle of radius $\sqrt{x}$, and thus travels a distance of $2\pi\sqrt{x}$. To put this in a form we can integrate, we use the ideas from the previous section to write

$$2\pi\sqrt{x}ds = 2\pi\sqrt{x}\sqrt{1+\left(\frac{dy}{dx}\right)^2}\,dx = 2\pi\sqrt{x}\sqrt{1+\frac{1}{4x}}dx = 2\pi\sqrt{x+\frac{1}{4}}dx = \pi\sqrt{4x+1}dx.$$

Since this represents the area of an infinitesimally thin frustum at a typical point x, the area of the entire surface is given by

$$\text{Surface Area} = \int_0^4 \pi\sqrt{4x+1}dx = \pi\left[\frac{1}{4}\cdot\frac{2}{3}(4x+1)^{3/2}\right]_0^4 = \frac{\pi}{6}\left[17\sqrt{17}-1\right].^{*} \quad \blacklozenge$$

The square root in the ds formula makes most integrals for arclength or surface area difficult or impossible to evaluate exactly. In such cases, we must settle for numerical approximations. Pragmatically speaking, this is no great loss, since a computer can approximate any definite integral to as many decimal places of accuracy as we need. You've encountered such a situation before when learning trigonometry: We can compute $\sin(60°)$ exactly, but not $\sin(61°)$.

---

* Those uncomfortable stepping across that second equals sign can always integrate by substitution instead: Let $u = 4x + 1$.

Exercises.

3. In parts a-b, find the surface areas of the following surfaces of revolution.

a) $y = x^3$ revolved around the x-axis, for $0 \leq x \leq 2$.

b) $y = x^2$ revolved around the $\boldsymbol{y}$-axis, for $0 \leq y \leq 4$. [*Hint: Recall exercise 2c.*]

c) The surface in part (b) has the same area as the surface in the example in the text. Why?

4. Everybody knows that a sphere of radius r has a surface area of $4\pi r^2$. Prove this by setting up and evaluating the appropriate integral for surface area.

5. Set up, but do not evaluate, integrals that represent the areas of the following surfaces.

a) $y = \sin x$ revolved around the x-axis, for $0 \leq x \leq \pi$.

b) $y = e^{-x}$ revolved around the x-axis, for $0 \leq x \leq 1$.

6. Figure out how to use a computer or calculator to approximate (to the nearest thousandth) the value of the integral that you found in Exercise 5a.

7. In Exercise 27 of Chapter 5, you showed that $\int \sec x\, dx = \ln|\sec x + \tan x| + C$.
A related integral that arises with surprising frequency is $\int \sec^3 x\, dx$.
How does one find the antiderivative of $\sec^3 x$? In this exercise, I'll point the way.

a) First, write $\int \sec^3 x\, dx = \int \sec x \sec^2 x\, dx$ and then integrate by parts.

b) In the resulting "outer integral", use a Pythagorean identity to rewrite the integrand entirely in terms of *secant*.

c) Use a linearity property to split this integral in two: an integral of secant, and an integral of secant cubed.

d) Since you already know an antiderivative of secant, you can insert this into your equation, and then solve for $\int \sec^3 x\, dx$ using the "self-replicating integral" trick from Exercise 42 of Chapter 5.

8. Use the result you found in the previous exercise to find the length of $y = x^2$ between the origin and $(1,1)$.
[*Hint: You'll need trig substitution and other trigonometric shenanigans.*]

9. In this exercise, you'll prove that a (full) cone's lateral surface area is its generator's length times the distance its midpoint travels while the cone is generated. Once we've proved this result, we'll be able to use it - in the very next exercise - to prove that the same statement holds for a conical *frustum*.

a) By cutting a cone along a generator and spreading it flat, we can convert it into a plane "pac-man", which is part of a circle. The ratio of the pac-man's area to the circle's area clearly equals the ratio of the round part of the pac-man's perimeter to the circle's circumference. Use this proportion to show that **the cone's lateral surface area is $\boldsymbol{\pi rs}$**, where s is the cone's "slant height" (i.e. the length of its generator) and r is the radius of its base.

b) Prove that $\boldsymbol{\pi rs}$ is also **(Cone's generator's length) x (Distance its midpoint travels)**. [*Hint: Similar triangles.*]

c) Combine the results of parts (a) and (b) to establish the theorem we wished to prove in this exercise.

10. A conical frustum's lateral surface area is clearly the difference of two cones' lateral surface areas. Using this idea, express a frustum's lateral surface area in terms of the radii of its bases and its slant height. Then express the product **(Frustum's generator's length) x (Distance its midpoint travels)** in terms of the same three quantities and verify that the result is the same, thus proving the frustum formula we've been using throughout this section.

[*Hint: This requires similar triangles and perseverance.*]

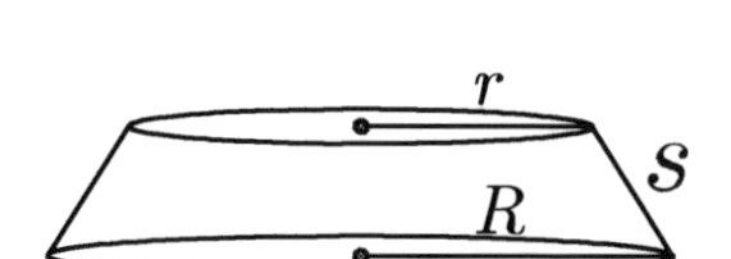

11. Since Chapter 4, we've been finding the volumes of solids of revolution by chopping them into cylindrical slabs. Yet now, when computing these solids' surface areas, we have forsaken the cylinders in favor of conical frusta. Why can't we use cylinders to find surface areas, too? The concise but unsatisfying answer is that cylinders simply don't work for surface areas. You'll demonstrate this now.

a) Without calculus, find the lateral surface area of a cone with height and base radius 1. [*Hint: Exercise* 9.]

b) Revolve the graph of $y = x$ (for $0 \leq x \leq 1$) around the x-axis to generate a cone of height and base radius 1. For the sake of argument, pretend that a valid method for finding its surface area is to cut it into infinitesimally thin *cylinders* and then to integrate the lateral surface areas of these cylinders. Carry this (invalid) procedure out and show that it does *not* yield the correct surface area, which you found in part (a).

12. (Optional - for those seeking a *deeper* answer to the question posed in the previous problem.)

We saw in the previous problem that cylinders won't work for surface area calculations, but what gives us the right to assume that conical frusta *will* work? To answer this question, we'll need to return to first principles. We'll also need to consider a subtle point that I deliberately elided back in Chapter 4.

Recall the fundamental premise upon which calculus is built: *On an infinitesimal scale, curves are straight.* Hence, we may think of a curve as a chain of infinitesimally short line segments, and the area beneath a curve as a set of infinitesimally thin *trapezoids*, one of which is shown in the first figure at right. (Note that one side of each trapezoid *coincides with the curve*, while another lies on the x-axis.)

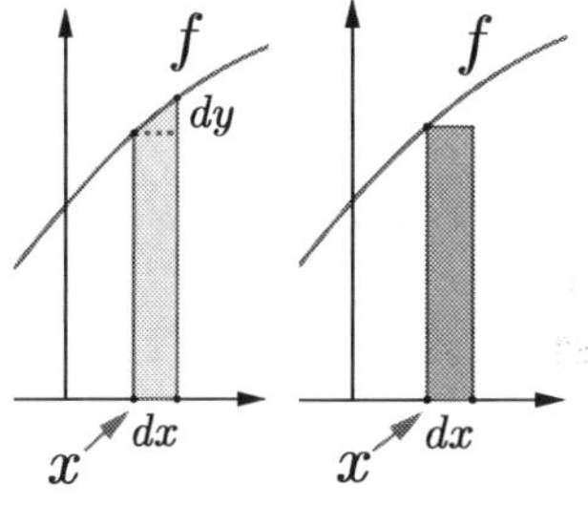

This should give you pause. Dissecting the area under a curve into trapezoids is obviously valid since the curve coincides with the trapezoids' tops, but... ever since Chapter 4's first page, we have been finding such areas with *rectangles*. Was this a mistake? The rectangles' horizontal tops generally do *not* coincide with the curve, so what gives us the right to use rectangles instead of trapezoids?

The answer, as you'll now discover, involves the second fundamental premise (from Chapter 1) upon which we built the infinitesimal calculus: *Second-order infinitesimals can be discarded as zeros.*

a) Explain why the infinitesimally-thin trapezoid and rectangle in the figure above have *the same area.*

> Part (a) justifies our use of rectangles when we compute areas. It also conveys an important insight: Because of the local straightness of curves, dissecting an area into trapezoids is obviously valid, but our ability to chop an area into *rectangles* is a gift of the gods for which we should be thankful!

b) Consider a solid of revolution. Can we find its **volume** by chopping it into infinitesimally thin conical frusta? How about its surface area? In each case, explain how you *know* that dissecting the solid into frusta will (or will not) yield the correct result.

c) Explain why cutting a solid of revolution into infinitesimally-thin cylindrical slabs (as opposed to conical frusta) yields, after integration, the correct volume of the solid. (Another gift of the gods!)

d) As you saw in the previous problem, the gods did *not* give us the gift of finding surface areas by means of cylinders. This, however, is no great loss. We have no right to expect such gifts. Meditate on this.

Improper Integrals

Under certain circumstances, the sum of infinitely many *real* numbers can be finite. You first encountered this phenomenon in elementary school (!), when you learned that the decimal 0.333... equals $1/3$. To expose the infinite sum hiding within this familiar fact, we need only re-express it as follows:

$$\frac{3}{10}+\frac{3}{100}+\frac{3}{1000}+\frac{3}{10{,}000}+\frac{3}{100{,}000}+\cdots=\frac{1}{3}.$$

Next, imagine continuing the graph at right (and its rectangles) down the full infinite extent of the positive x-axis. Since the rectangles' areas correspond to terms in the sum above, the sum of all of their areas must, according to your elementary school teacher, be $1/3$. Moreover, since the area under the curve (for $x \geq 1$) is clearly less than the total area of all the rectangles, it follows that

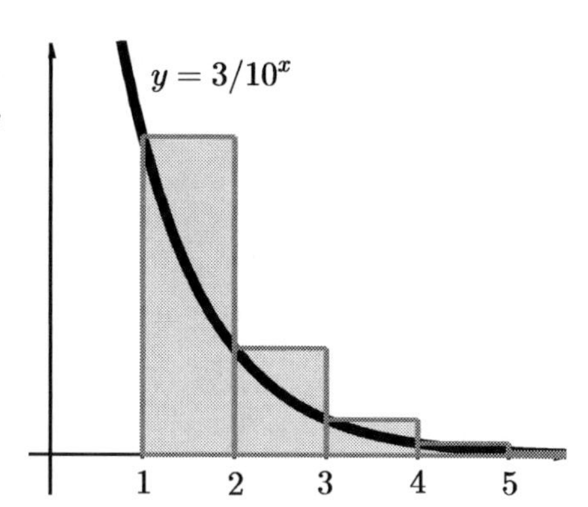

$$\int_1^\infty \frac{3}{10^x}dx < 1/3.$$

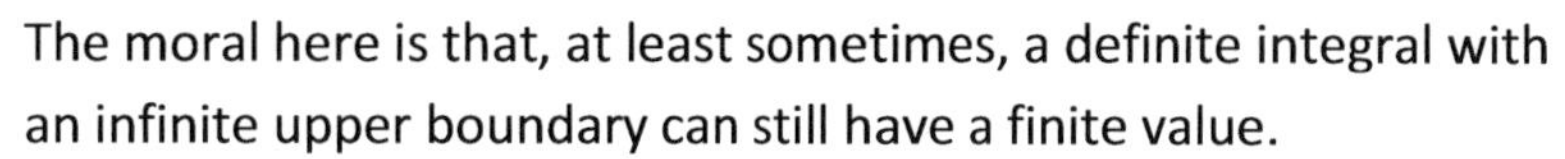

The moral here is that, at least sometimes, a definite integral with an infinite upper boundary can still have a finite value.

In fact, we can find the exact value of the so-called "improper" integral we've just written down. To integrate from a to ∞, the trick is simple: Integrate from a to b (where b is just a symbolic placeholder), and then take a limit as the upper boundary b goes to infinity.

Example 1. Evaluate the integral $\int_1^\infty \frac{3}{10^x}dx$.

Solution. To evaluate this improper integral, we'll express it as the limit of a proper integral:

$$\int_1^\infty \frac{3}{10^x}dx = \lim_{b\to\infty}\int_1^b \frac{3}{10^x}dx. \qquad \bigstar$$

We'll return to the limit in a moment. For now, let us concentrate on the proper integral. If we recall that $(10^x)' = (\ln 10)10^x$, we can integrate by guessing and adjusting:

$$\int_1^b \frac{3}{10^x}dx = 3\int_1^b 10^{-x}\,dx = 3\left[\frac{-1}{\ln 10}10^{-x}\right]_1^b = -\frac{3}{\ln 10}\left(10^{-b}-10^{-1}\right).$$

This, combined with equation ($\bigstar$), yields

$$\int_1^\infty \frac{3}{10^x}dx = \lim_{b\to\infty}\int_1^b \frac{3}{10^x}dx = \lim_{b\to\infty}\left[-\frac{3}{\ln 10}\left(10^{-b}-10^{-1}\right)\right].$$

As $b\to\infty$, the expression 10^{-b} vanishes. Nothing else in the brackets depends on b, so we have

$$\int_1^\infty \frac{3}{10^x}dx = \lim_{b\to\infty}\left[-\frac{3}{\ln 10}\left(10^{-b}-10^{-1}\right)\right] = -\frac{3}{\ln 10}\left[0-\frac{1}{10}\right]$$

$$= \frac{3}{(10\ln 10)} \approx \mathbf{0.130}. \qquad \blacklozenge$$

In brief, an improper integral's value is the limit of a related proper integral. If the limit exists (and is finite), we say that the integral *converges*. If the limit doesn't exist (or if it "blows up to infinity"), we say that the integral *diverges*, meaning that it lacks a finite limiting numerical value.

Example 2. Explain why the improper integrals $\int_0^\infty x^2 dx$ and $\int_0^\infty \cos x\, dx$ are divergent.

Solution. By definition,

$$\int_0^\infty x^2 dx = \lim_{b\to\infty} \int_0^b x^2 dx = \lim_{b\to\infty} b^3/3 = \infty,$$

so this integral diverges because the related limit blows up. This, I hope, is geometrically obvious; the area under $y = x^2$ for $0 \le x \le b$ clearly gets bigger and bigger as b goes to infinity. In contrast,

$$\int_0^\infty \cos x\, dx = \lim_{b\to\infty} \int_0^b \cos x\, dx = \lim_{b\to\infty} (\sin b),$$

so this integral diverges because the related limit doesn't exist: Sine oscillates between -1 and 1 as its argument goes to infinity, so it will never approach a limiting value. ♦

If an improper integral is to have any hope of converging, the graph of its integrand clearly must vanish (i.e. approach zero) as x goes to ∞. A vanishing integrand, however, is not enough to ensure convergence. The integrand must not only vanish; it must vanish *quickly*. Consider the graphs of $y = 1/x$ and $y = 1/x^2$. Both vanish, but $1/x^2$ vanishes *faster*. As you'll see in the exercises, the integral of $1/x$ from 1 to ∞ diverges, while that of $1/x^2$ converges. Differences in "vanishing speed" can thus mean the difference between a finite area and an infinite area – a difference not of degree, but of kind.

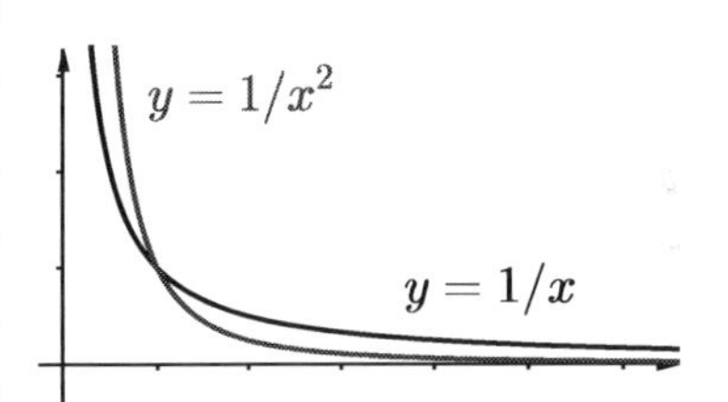

Finally, a word on notation. Although we define improper integrals as limits of proper integrals, we rarely write the limits out in practice. It is more common and convenient to avoid clutter by condoning a slight abuse of notation and writing, for example,

$$\int_1^\infty \frac{1}{x^3} dx = \left[\frac{1}{-2x^2}\right]_1^\infty = \left[0 - \left(-\frac{1}{2}\right)\right] = \frac{1}{2},$$

where we understand "evaluating a function at ∞" to mean *taking a limit of the function as $x \to \infty$.*

Exercises.

13. a) Find the value to which $\int_1^\infty 1/x^2\, dx$ converges. b) Show that $\int_1^\infty 1/x\, dx$ diverges.

c) Find the smallest integer m for which $\int_1^m 1/x\, dx > 1$.

For this value of m, $\int_1^m 1/x^2\, dx$ remains *less than* 1, as part (a) suggests. Find its value.

d) Find the smallest integer n for which $\int_1^n 1/x\, dx > 20$. For this value of n, find $\int_1^n 1/x^2\, dx$ to 9 decimal places.

14. Determine whether the following improper integrals converge or diverge. Find the values of those that converge.

a) $\int_0^\infty e^{-x}\, dx$ b) $\int_1^\infty \ln x\, dx$ c) $\int_1^\infty \frac{1}{x^2}\sin\left(\frac{1}{x}\right) dx$ d) $\int_0^\infty \frac{x^2}{1+x^6} dx$

15. Just as an integral's upper boundary of integration can be ∞, its *lower* boundary of integration can be $-\infty$.

a) Explain how to rewrite the improper integral $\int_{-\infty}^{b} f(x)dx$ as the limit of a proper integral.

b) Integrals over the entire real line (from $-\infty$ to ∞) are common in probability theory. Find

$$\int_{-\infty}^{\infty} \frac{1}{1+x^2}dx.$$

Also, sketch the graph of the function being integrated, and interpret the integral in terms of area.

c) The function $f(x) = e^{-x^2}$ lies at the heart of the so-called "normal distribution" in probability theory. Alas, it has no elementary antiderivative. Nonetheless, mathematicians have proved (without the FTC!) that

$$\int_{-\infty}^{\infty} e^{-x^2}dx = \sqrt{\pi}.$$

Your problems: First, take a minute to admire this striking relationship between e and π (by way of infinity). Second, sketch the graph of the function being integrated, and interpret the integral in terms of area.

d) Tweaking the function $y = e^{-x^2}$ a bit yields the "probability density function" for a normal random variable with mean μ and standard deviation σ. [*If this means nothing to you, don't worry; you can still do the problem that follows.*] To obtain this jewel of probability, we must stretch and shift $y = e^{-x^2}$ in such a way that the resulting bell-shaped graph satisfies the following three conditions:

1. The bell must be symmetric about the line $x = \mu$.
2. The x-coordinates of the bell's inflection points must be $\mu \pm \sigma$.
3. The total area under the graph must be exactly 1.

Your problem: **Find the function we seek.**

16. Revolving the curve $y = 1/x$ (for $x \geq 1$) around the x-axis generates the solid known as *Gabriel's horn*.

a) Is the volume of Gabriel's horn finite or infinite?

b) Set up an integral for the surface area of Gabriel's horn. Use the obvious fact that $1 + (1/x^4) > 1$ for all x to show that the horn's surface area exceeds an area that you've already shown is infinite (in exercise 13b). From this, conclude that Gabriel's horn has an infinite surface area.

c) An old saying regarding Gabriel's horn: "You can fill it with paint, but you can't paint it." Meditate on this.

17. The *gamma function* is a non-elementary function defined as follows: $\Gamma(t) = \int_0^{\infty} e^{-x}x^{t-1}dx$.

a) Find $\Gamma(1)$ and $\Gamma(2)$. b) Use integration by parts to prove that $\Gamma(n+1) = n\Gamma(n)$ for all whole numbers n.*

c) Use your results from parts (a) and (b) to find $\Gamma(3)$, $\Gamma(4)$, $\Gamma(5)$, and so forth until you see the connection with the factorial function.

18. The graphs of $y = 1/x^r$ (for positive values of r) vanish at different rates, sometimes rapidly enough to make the area under the curve (for $x > 1$) finite, but sometimes not. The boundary between these cases turns out to be $r = 1$. Demonstrate that the area is finite if $r > 1$, but infinite if $r \leq 1$.

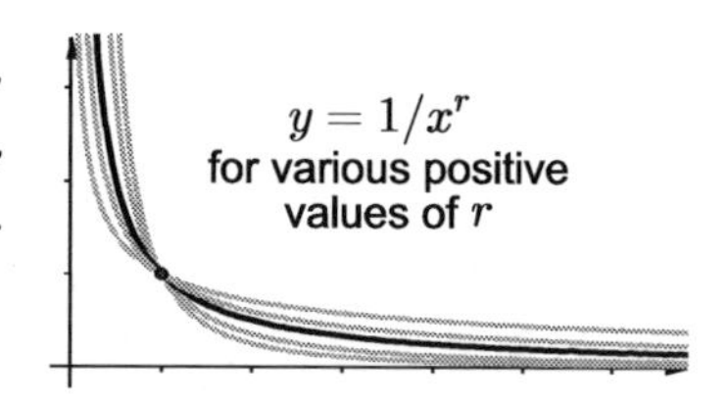

---

* A simple fact that will help you in this problem: Applying IBP to a *definite* integral yields $[\text{top product}]_a^b - \int_a^b(\text{outer integral})$. That is, you can evaluate the top product numerically even if you can't solve the outer integral.

Riemann Sums

In earlier chapters, you learned that we can consider a derivative either as a ratio of infinitesimals (dy/dx) or as the *limit* of a ratio of real numbers (namely, the limit of $\Delta y/\Delta x$ as the denominator vanishes). Although working with infinitesimals is more direct, some people find them so ontologically disturbing that they prefer the prophylactic safety of the limits, which takes them as close as possible to infinitesimals without ever making actual contact.

In this section, we'll discuss the limit approach to definite integrals. Understanding integrals entirely in terms of real numbers (i.e. without reference to infinitesimals) isn't just a pedantic intellectual exercise. It will help you understand how a computer can *approximate* a definite integral's value. In fact, the best way to understand the integral's limit definition is to begin with the humbler problem of approximation.

The function at right lacks an elementary antiderivative, so we cannot use the FTC to integrate it over an interval (say, $[1,3]$). We can, however, approximate the integral's value: Cut the interval $[1,3]$ into 10 equal pieces and construct rectangles on the pieces; the sum of their areas approximates the area under the curve. With patience and a pocket calculator, we can therefore determine that

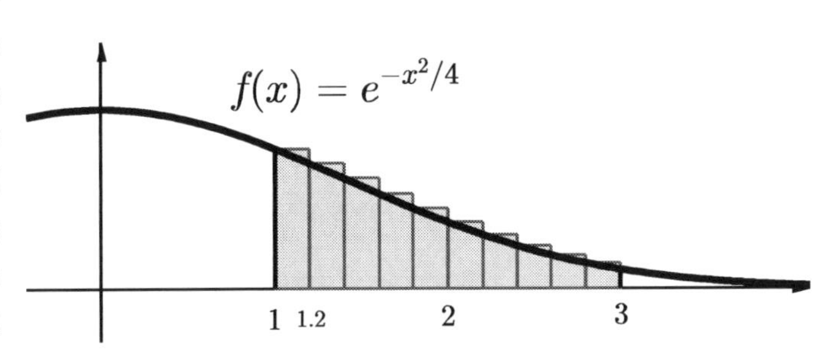

$$\int_1^3 e^{-x^2/4}dx \approx e^{-1^2/4}(0.2) + e^{-(1.2)^2/4}(0.2) + \cdots + e^{-(2.8)^2/4}(0.2) \approx 0.8579.$$

This approximation, of course, is quite crude. (The integral's value, to four decimal places, is actually 0.7898.) To improve it, we could use, say, 40 rectangles. Because these rectangles will be only a quarter as wide as before, they will hug the curve better, creating a better approximation:

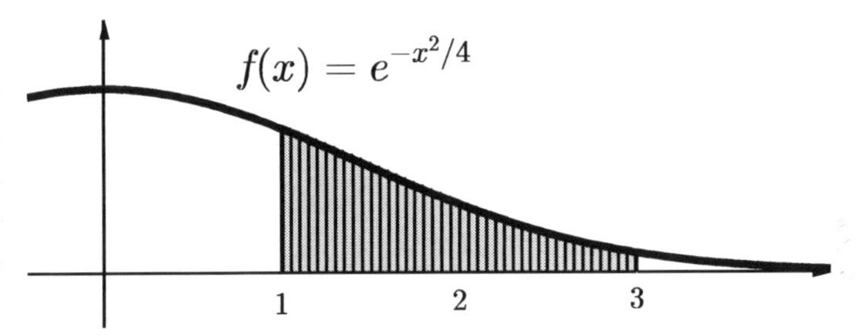

$$\int_1^3 e^{-x^2/4}dx \approx e^{-1^2/4}(0.05) + e^{-(1.05)^2/4}(0.05) + \cdots + e^{-(2.95)^2/4}(0.05) \approx 0.8067.$$

Naturally, no one does such calculations by hand. This is a job for a computer. Still, it is good to know that this is essentially the type of calculation that the computer is doing. With 1000 rectangles, our approximation would be 0.7905, and with 5000 rectangles, it would be 0.7898. When you ask a computer program for an integral's numerical value, it always uses enough terms in the sum to guarantee that its approximation will be accurate to as many decimal places as it displays on the screen.

The sums above are best thought of as *weighted sums*: After cutting the interval $[1,3]$ into equal pieces, we multiplied each piece's length by a certain *weight*, and then summed the results. The weights, of course, were the rectangles' heights, which we found by evaluating the integrand at each subinterval's left endpoint.[*] Cutting the interval into more – and tinier – pieces would improve the approximation. The extreme case of cutting it into infinitely many infinitesimal pieces would make it exact.

Pure souls who wish to avoid saying "the i-word" can recast that last sentence euphemistically: The approximation becomes exact *in the limit* as the number of pieces in our weighted sum goes to infinity (or, to put it even more delicately, as the pieces' common length goes to zero). Therefore, if we wish to define a definite integral in purely real terms (without reference to infinitesimals), we begin by recognizing an integral as *the limit of a weighted sum*.

To capture this last idea in symbols, we must first consider how we obtained the weighted sum itself. We began by cutting the interval $[a, b]$ into n subintervals with a common length of $\Delta x = (b - a)/n$. We then picked a "test point" in each subinterval (say, the left endpoint), and used the values of f at the test points as the weights by which we multiplied each subinterval's length. Hence, if we call our test points $x_1, x_2, \ldots, x_n$, then our weighted sum (our approximation of the integral's value) will be

$$f(x_1)\Delta x + f(x_2)\Delta x + \cdots f(x_n)\Delta x.$$

As discussed above, this approximation becomes exact in the limit as $n \to \infty$. Or expressed symbolically,

$$\int_a^b f(x)dx = \lim_{n \to \infty} [f(x_1)\Delta x + f(x_2)\Delta x + \cdots f(x_n)\Delta x].$$

Using *sigma summation notation*, we can rewrite this last statement more compactly as

$$\int_a^b f(x)dx = \lim_{n \to \infty} \left[\sum_{i=1}^{n} f(x_i)\, \Delta x \right].^{\dagger}$$

Limit-based treatments of calculus take the preceding equation as the definite integral's *definition*, and then develop the theory of integration (the FTC and so forth) from it. It is possible, but exceedingly tedious, to evaluate certain (very simple) definite integrals directly from this "limit recipe". I'll not bore you with the details. The weighted sum itself is often called a **Riemann sum**, after the great 19th-century mathematician Bernhard Riemann. One may summarize the limit-based approach to calculus as follows: The derivative is the limit of a difference quotient; the definite integral is the limit of a Riemann sum.

[*] We could just as easily have used the subintervals' *right* endpoints. Using different "test points" to obtain the weights (left endpoints, right endpoints, midpoints, or whatnot) yields slightly different approximations, as you'll see in exercise 23. Had we used right endpoints, for example, our sums would have underestimated the integral's value, instead of overestimating it. You can understand why by looking at the graph and thinking about what the rectangles would have looked like in that case.

[†] If sigma summation notation is new to you, fear not. It's just shorthand. The capital sigma (Σ) tells us to sum all instances of the expression that follows it as the index i runs through all integers between its stated boundaries (here, from 1 to n). For example, $\sum_{i=2}^{5} i^2$ is shorthand for $2^2 + 3^2 + 4^2 + 5^2$. Similarly, $\sum_{i=6}^{n} x_i$ just means $x_6 + x_7 + x_8 + \cdots + x_n$.

Exercises.

19. Evaluate the following expressions involving sigma summation notation.

a) $\sum_{i=7}^{10} i$ b) $\sum_{i=0}^{5}(3i + 2^i)$ c) $\sum_{i=100}^{103} \sin\left(\frac{\pi}{2} i\right)$

20. Express the following sums in sigma summation notation.

a) $1 + 2 + 3 + \cdots + 100$ b) $\frac{1}{4} + \frac{1}{9} + \frac{1}{16} + \cdots + \frac{1}{400}$ c) $1 - \frac{1}{2} + \frac{1}{6} - \frac{1}{24} + \frac{1}{120} - \frac{1}{720} + \frac{1}{5040}$

21. The sum in exercise 20a occupies a special place in mathematical folklore. Late in the 18th century, a harried German school teacher sought to settle his rambunctious class of children by making them find the sum of the first hundred numbers. Within seconds, one boy announced the answer: 5050. The astonished teacher asked the boy how he had determined this. He explained that he had imagined the numbers written backwards and forwards in the following pattern.

$$\begin{matrix} 1 & 2 & 3 & \cdots & 98 & 99 & 100 \\ 100 & 99 & 98 & \cdots & 3 & 2 & 1 \end{matrix}$$

The numbers in each column, he noted, add up to 101, and there are 100 columns, so the total of all the numbers in the two rows is 10100. Since the rows contain the exact same numbers, the sum of those in the *first* row must be exactly half of 10100, which is 5050.

Little Karl Friedrich Gauss went on to become one of the greatest mathematicians who ever lived.

Use Baby Gauss's idea to find a formula for the sum of the first n whole numbers. Once you've found it, express the statement of the formula using sigma summation notation.

22. Certain numbers of pebbles can be arranged into square patterns of the sort shown in the figure at right. These numbers $(1, 4, 9, 16, 25, \ldots)$ are called the *square numbers*. For the purposes of this exercise, we'll denote the nth square number by the symbol $\square_n$.

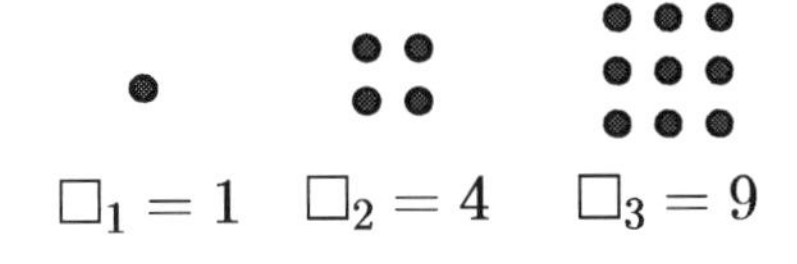

Considerably less well-known are the *triangular numbers*, the first three of which are shown at right. We'll denote the nth triangular number by the symbol Δ_n. Some problems for your consideration…

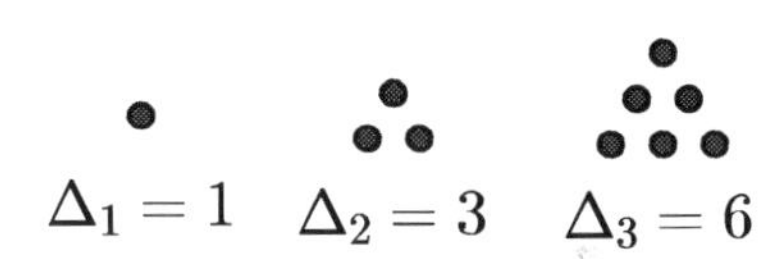

a) Find formulas for $\square_n$ and Δ_n in terms of n. [*Hint: Baby Gauss.*]

b) Besides 1, are any numbers both square and triangular?

c) The sum of consecutive triangular numbers is always a square number (e.g. $3 + 6 = 9$.) Explain why this is so.

d) The *perfect numbers* are those equal to the sum of their proper divisors (i.e. divisors smaller than themselves). As of December 2018, exactly 50 perfect numbers have been discovered. The first perfect number is 6, since it is the sum of its proper divisors (1, 2, and 3). Find the second perfect number.

e) A *Mersenne prime* is a prime of the form $2^n - 1$. (A better name would be "Euclidean prime" since Euclid studied them 2000 years before Mersenne, but such is life.) As of December 2018, exactly 50 Mersenne primes have been discovered. The first Mersenne prime is $3 = 2^2 - 1$. Find the next three.

f) Euclid proved that if M is any Mersenne prime, then Δ_M is perfect. Verify this in a few cases.

23. (Arithmetic Ho!) To develop a feel for Riemann sums, use them to *approximate* the value of $\int_0^1 x^2\, dx$ by…

a) Cutting the interval $[0,1]$ into 10 pieces and using their left endpoints to determine the weights in the sum.

b) Cutting the interval $[0,1]$ into 10 pieces and using their right endpoints to determine the weights.

c) Cutting the interval $[0,1]$ into 10 pieces and using their midpoints to determine the weights.

d) Which of the preceding methods produces the best approximation to the true value of the integral? Can you suggest a plausible geometric explanation as to why it offers the best approximation?

Escape From Flatland

"Either this is madness or it is Hell."
"It is neither," calmly replied the voice of the Sphere, "it is Knowledge; it is Three Dimensions: open your eye once again and try to look steadily."

- Edwin A. Abbott, *Flatland: A Romance of Many Dimensions*, Chapter 18.

In this section, we shall endeavor to make sense of the exotic idea of a *hypersphere* in four-dimensional space, at least to such an extent that we will be able to derive a formula its (hyper)volume. Before we can cope with four-dimensional space, however, we'll need to acquaint ourselves with some basic coordinate geometry in the ordinary space of three dimensions.

Setting up three mutually perpendicular axes affords us the luxury of fusing (or confusing?) points in space with ordered triples of real numbers: Each point corresponds to an ordered triple, and vice-versa. Coordinates such as $(4, -2, 1)$ tell us how to reach this point from the origin: Go 4 units along the positive x-axis, then 2 in the direction of the negative y-axis, then 1 in the direction of the positive z-axis.

Among the many ways to move from a point A to another point B in space, we can always take the scenic route along three roads, each parallel to an axis. On this route, a point changes its coordinates one at a time. The three roads have lengths of $|\Delta x|, |\Delta y|$, and $|\Delta z|$, and meet at right angles, so we can apply the Pythagorean Theorem twice (first to ΔAB_1B_2, then to ΔAB_2B) to find AB, thus establishing the three-dimensional distance formula:

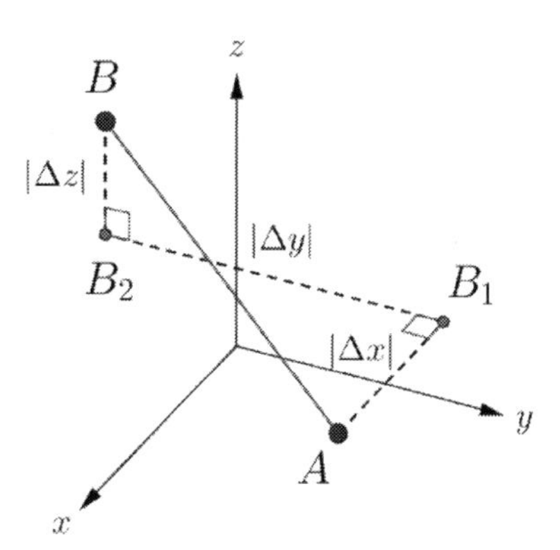

$$AB = \sqrt{(\Delta x)^2 + (\Delta y)^2 + (\Delta z)^2}.$$

Note that the 3-D distance formula has the same form as its 2-D parent; its only novelty is a third term under the radical.

The distance formula helps us derive the equations of surfaces in space. A sphere, for example, is the set of all points at a fixed distance r from a given point (a, b, c). Thus, by the distance formula, a point (x, y, z) lies on the sphere if and only if $\sqrt{(x-a)^2 + (y-b)^2 + (z-c)^2} = r$. Consequently, this is the sphere's equation. We usually write it, however, in the cleaner form that results from squaring both sides.

> The equation of a sphere of radius r centered at (a, b, c) is
>
> $$\boldsymbol{(x-a)^2 + (y-b)^2 + (z-c)^2 = r^2}.$$

The sphere's equation is thus just like the circle's, but with a third term for the third dimension.

The equation of the xy-plane (i.e. the plane containing the x and y axes) is $z = 0$, since a point lies in it if and only if its z-coordinate is 0. Planes parallel to the xy-plane have equations of the form $z = k$. It is geometrically obvious that a sphere $x^2 + y^2 + z^2 = r^2$ and a plane $z = k$ (where $|k| < r$) intersect in a circle, but we can see this algebraically, too: Substituting the plane's equation into the sphere's yields $x^2 + y^2 = r^2 - k^2$, the equation of a circle. More particularly, this shows that the intersection of the sphere $x^2 + y^2 + z^2 = r^2$ and a plane k units above the xy-plane is a circle of radius $\sqrt{r^2 - k^2}$.

There are many ways to derive the formula for a sphere's volume, some of which you've seen already. Here's a way that uses three-dimensional coordinate geometry – and generalizes to higher dimensions.

Claim. A sphere of radius r has volume $(4/3)\pi r^3$.

Proof. Centered at the origin, its equation is $x^2 + y^2 + z^2 = r^2$. Think of the sphere as a stack of infinitesimally thin discs. Consider a typical disc z units above the xy-plane. By our work above, its radius is $\sqrt{r^2 - z^2}$, so the area of its circular base is $\pi(r^2 - z^2)$. Multiplying this area by the disc's infinitesimal thickness gives us the disc's volume: $\pi(r^2 - z^2)dz$.

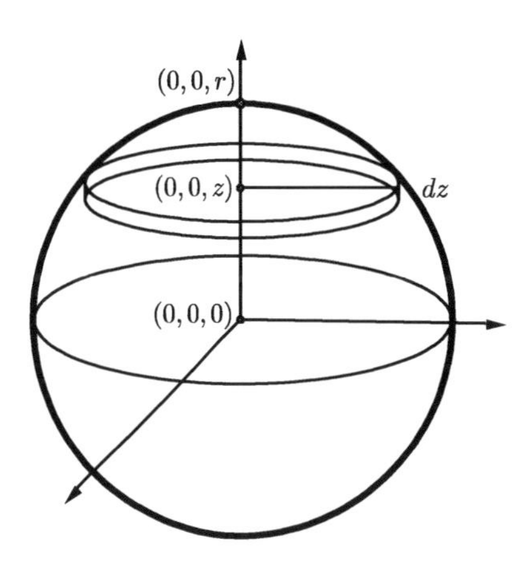

The sphere's volume is thus

$$2\int_0^r \pi(r^2 - z^2)dz = 2\pi\left[r^2 z - \frac{z^3}{3}\right]_0^r = \frac{4}{3}\pi r^3. \quad \blacksquare$$

Edwin Abbott's novella *Flatland*, published in 1884, describes a civilization of two-dimensional beings for whom the idea of a third spatial dimension is inconceivable. To his distress, the book's narrator is visited by an intelligent sphere, who manifests himself to the narrator by passing through the plane of Flatland. The narrator, however, sees only a sequence of two-dimensional cross-sections of the sphere: circles whose radii grow and then shrink. Lacking experience of a third spatial dimension, the narrator can't mentally assemble these cross-sections to envision the sphere as it truly is. Eventually, the sphere rips the narrator bodily out of Flatland and shows him Spaceland. When the newly-enlightened narrator asks the sphere to show him the *fourth* dimension, too, the irritated sphere flings him back down to Flatland, dismissing the idea of a fourth spatial dimension as inconceivable nonsense.

Four-dimensional space admits *four* mutually perpendicular axes (x, y, z, w), which we can erect at any point. A Flatlander could pick a point to be the origin, confidently set up the x and y axes, and then have no idea where a z-axis should go. We Spacelanders laugh at his confusion, draw the "inconceivable" z-axis… and then find ourselves in the same confusion as the Flatlander, but with respect to the w-axis.

The z-axis intersects Flatland at a single point, the origin. From the Flatlanders' perspective, the rest of the z-axis lies in some mysterious "elsewhere". Similarly, the w-axis intersects Spaceland at the origin, but the rest of it lies elsewhere. Try (and fail) to picture this; you'll feel sympathy for the Flatlanders.

Points in four-dimensional space have four coordinates. Coordinates such as $(4, -2, 1, 3)$ tell us how to reach this point from the origin: Move 4 units along the positive x-axis, then 2 in the direction of the negative y-axis, then 1 in the direction of the positive z-axis, which takes us out of Flatland, and finally, we move 3 units in the direction of the positive w-axis, which takes us out of Spaceland. Clearly, a point lies in Spaceland if and only if its w-coordinate is zero. Consequently, the equation of Spaceland itself, relative to the four-dimensional space we are describing, is $w = 0$.

Flatland is one of many planes in Spaceland, which in turn is one of many *hyperplanes* in four-dimensional space, each extending infinitely in three dimensions, but not at all in a fourth. Moving one of Spaceland's points a unit "outwards" in the w-direction would shift it to a parallel hyperplane, $w = 1$. Naturally, all hyperplanes parallel to $w = 0$ have equations of the form $w = k$, where k is a constant.

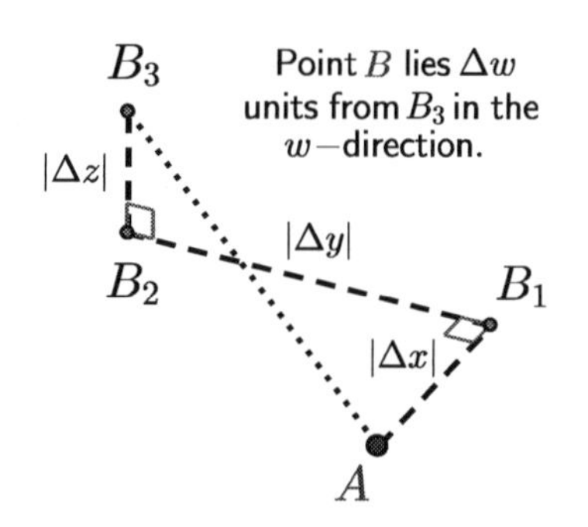

You can surely guess the four-dimensional distance formula, but deriving it carefully offers good practice grappling with four-dimensional space. (And besides, guesses are often wrong.) As in our derivation of the three-dimensional distance formula, we'll move from point A to point B in four-dimensional space via the "scenic route" along four roads, each parallel to an axis. Since one coordinate changes per road, the roads' lengths are $|\Delta x|, |\Delta y|, |\Delta z|$, and $|\Delta w|$. Clearly, we can't draw the journey's last leg (from B_3 to B) in the same picture, but this need not deter us.

Applying the Pythagorean Theorem to ΔAB_1B_2 and ΔAB_2B_3 yields $AB_3 = \sqrt{(\Delta x)^2 + (\Delta y)^2 + (\Delta z)^2}$. To link AB_3 to AB (the distance we seek), we consider ΔAB_3B. We can only see one side of this triangle in our figure, but it is a perfectly ordinary triangle from a four-dimensional being's perspective. (It may help to compare the analogous situation for a Flatlander contemplating a triangle whose base lies in Flatland, but whose third vertex doesn't.)

Any two points determine a 1-dimensional space, a line. Any three points (provided they are not on the same line) determine a 2-dimensional space, a plane. Any four points (provided they are not all in the same plane) determine a 3-dimensional space, a hyperplane. As we proceed along our scenic route from A to B, each successive road is perpendicular to the whole space determined by the points we've already encountered on the scenic route. For instance, road B_1B_2 is perpendicular to *line* AB_1, and road B_2B_3 is perpendicular to *plane* AB_1B_2. Similarly, road B_3B is perpendicular to *hyperplane* $AB_1B_2B_3$. This means, more specifically, that B_3B is perpendicular to any line inside that hyperplane that passes through B_3. In particular, it means that B_3B is perpendicular to AB_3.*

Since lines B_3B and AB_3 are perpendicular, ΔAB_3B is a *right* triangle. We know the lengths of its legs (We found AB_3 above, and $BB_3 = |\Delta w|$), so the Pythagorean Theorem will give us its hypotenuse, and with it, our four-dimensional distance formula:

$$\boldsymbol{AB = \sqrt{(\Delta x)^2 + (\Delta y)^2 + (\Delta z)^2 + (\Delta w)^2}}.$$

A *hypersphere* in four-dimensional space is defined as the set of all points at some fixed distance r (the hypersphere's radius) from a given point (a, b, c, d), the hypersphere's center. Though we can't picture a hypersphere in our poor three-dimensional minds, we can still deduce various things about it. Translating its definition into symbols (with our new distance formula), for instance, we see that a point (x, y, z, w) lies on the hypersphere if and only if $\sqrt{(x-a)^2 + (y-b)^2 + (z-c)^2 + (w-d)^2} = r$. This, the hypersphere's equation, looks cleaner if we square both sides:

In 4-D space, the equation of a hypersphere of radius r centered at (a, b, c, d) is

$$\boldsymbol{(x-a)^2 + (y-b)^2 + (z-c)^2 + (w-d)^2 = r^2}.$$

* Those comfortable with vectors may appreciate this quick proof that $B_3B \perp AB_3$: First, $\overrightarrow{B_3B} = \langle 0,0,0,d\rangle$ for some constant d, since on road B_3B, only the w-coordinate changes. For similar reasons, $\overrightarrow{B_3A} = \langle a, b, c, 0\rangle$ for some constants a, b, and c. The dot product of these vectors is zero, so they are perpendicular, as are the lines containing them, B_3B and AB_3.

We now have enough information to determine any hypersphere's *hypervolume*.* The derivation will be very much in the spirit of our earlier proof of the formula for the sphere's volume. This time, however, we can't draw a picture to accompany it.

Problem. Derive a formula for the hypervolume of a hypersphere of radius r.
Solution. Centered at the origin, our hypersphere's equation will be $x^2 + y^2 + z^2 + w^2 = r^2$. We'll find its volume by slicing it into a stack of cross-sections. These cross-sections are basically the hypersphere's intersections with the hyperplanes $w = k$ (where $|k| \leq r$). We must remember, however, that the three-dimensional cross-sections are "stacked up" in the fourth dimension along the w-axis (so as to form a four-dimensional object, the hypersphere). Consequently, each cross-sectional slice is endowed with an infinitesimal thickness dw.

Slicing the hypersphere with a hyperplane k units up the w-axis (i.e. the hyperplane $w = k$) yields an object whose equation is $x^2 + y^2 + z^2 = r^2 - k^2$, which we recognize as a sphere of radius $(r^2 - k^2)^{1/2}$, and hence of volume $(4/3)\pi(r^2 - k^2)^{3/2}$.

Thus, a typical slice taken w units up the w-axis yields a sphere of volume $(4/3)\pi(r^2 - w^2)^{3/2}$; combined with its infinitesimal thickness dw, it contributes $(4/3)\pi(r^2 - w^2)^{3/2}dw$ to the hypersphere's hypervolume. To obtain the hypersphere's full hypervolume, we simply integrate the preceding expression as w runs from $-r$ to r. Or better still, since the integrand is an *even* function of w, we can integrate from 0 to r and double the result. (Recall Exercise 11 in Chapter 4.) Thus, the hypervolume we seek is

$$2\int_0^r (4/3)\pi(r^2 - w^2)^{3/2}dw = \frac{8}{3}\pi \int_0^r (r^2 - w^2)^{3/2}dw = \cdots = \frac{\pi^2}{2}r^4.^{\dagger} \quad \blacklozenge$$

Thus, to sum up...

> In four-dimensional space, a hypersphere of radius r has a hypervolume of $\boldsymbol{\frac{\pi^2}{2}r^4}$.

The presence of r^4 and π in this formula was to be expected; hypervolume is a four-dimensional quantity, and a hypersphere is a "round" object. But why *two* factors of π? I leave it to you to contemplate this mystery. When you've learned some multivariable calculus, you'll be able to derive a more general result: The formula for the n-dimensional hypervolume of a hypersphere in n-dimensional space turns out to be

$$\frac{\pi^{n/2}}{\frac{n}{2}\Gamma\left(\frac{n}{2}\right)}r^n.$$

(And yes, that is the *gamma function* in the denominator, which you met in exercise 17.)

* Hypervolume is measured in units^4. Just as a unit^2 of area is the content inside a "unit square" (whose 4 edges have unit length), and a unit^3 of area is the content inside a "unit cube" (whose 12 edges have unit length), a unit^4 of hypervolume is the content inside a "unit hypercube" (whose **32** edges all have unit length).

† I'll leave the integration details to you. First step: Let $w = r\sin\theta$.

Exercises.

24. During a 4-D baseball game played with a large hyperspherical baseball, a foul ball crashes through our humble three-dimensional Spaceland. You have the privilege of witnessing the passage of this extradimensional comet.

a) Describe what you see as the ball passes through our space.

b) The ball's radius is 2 feet. What is its hypervolume?

25. Describe the intersection of the hyperspheres $x^2 + y^2 + z^2 + w^2 = 1$ and $(x-1)^2 + y^2 + z^2 + w^2 = 1$.

26. Flatlanders would be disturbed not only by the exotic objects, such as spheres, within three-dimensional space, but also by three-dimensional space itself, since its roominess enables even familiar objects, such as lines, to interact in ways that are inconceivable in two dimensions. For instance, Flatlanders "know" that any pair of nonparallel lines must intersect. In three dimensions, however, nonparallel, nonintersecting lines can exist.

a) Give an example to explain how such lines (called *skew lines*) can exist in space.

b) If we pick two lines in space at random, they will almost certainly be skew to one another. Explain why.

c) In Spaceland, everybody knows that distinct planes can intersect only in a line. In four-dimensional space, however, two-dimensional planes can intersect in a single *point*. For example, the xy-plane (i.e. the plane containing the x and y axes) and the zw-plane (defined analogously) intersect this way. Justify my assertion that these planes intersect in a single point .

27. A two-dimensional object has both area (its 2D "inner content") and perimeter (its 1D boundary's length). A three-dimensional object has both volume (its 3D "inner content") and surface area (its 2D boundary's area). A four-dimensional object has both hypervolume (its 4D "inner content") and **boundary content** (its 3D boundary's volume). Find a formula for the boundary content of a 4D hypersphere of radius r.

[*Hint: This is easier than you might think. How are the formulas for a circle's area and circumference related?*]

28. Naturally, we can consider five-dimensional geometry, too. For this exercise, we'll call the 5th axis the u-axis.

a) What is the equation of the five-dimensional hypersphere of radius r centered at the origin?

b) The graph of $u = 0$ is a 4D hyperplane inside 5D space. The 4D hyperplanes parallel to it have equations of the form $u = k$. Describe the intersection of such a hyperplane with the hypersphere in part (a).

c) Find the 5D "inner content" of the 5D hypersphere of radius r.

d) Find the 4D boundary content of the 5D hypersphere of radius r. [*Compare exercise 27.*]

29. a) If we inscribe a circle in a square, what percentage of the square's area lies inside of the circle?

b) If we inscribe a sphere in a cube, what percentage of the cube's volume lies inside of the sphere?

c) Same question, but with a 4D hypersphere inscribed a 4D hypercube.

d) Same question, but with a 5D hypersphere inscribed a 5D hypercube. [*You'll need Exercise 28c for this.*]

e) Can you guess what happens to this percentage as the number of dimensions increases?

Part III
Special Topics

Chapter 7
Taylor Polynomials and Series

Taylor Polynomials

Press a few buttons and your calculator will inform you matter-of-factly that the sine of half a radian is about 0.479425539. This value isn't stored in your calculator's memory, so where did it come from?

Calculators are masters of arithmetic – and of nothing else.* Since evaluating a polynomial is a purely arithmetic task, a calculator can do it at lightning speed. The sine function, however, is no polynomial: It is defined *geometrically*. Consequently, your calculator has no conception of what $\sin(0.5)$ even means. Still, its programmer has taught it how to bluff. If, for instance, we ask a calculator to evaluate $\sin(0.5)$, it might pretend that we've asked it to evaluate the following 15th-degree *polynomial* at 0.5:

$$y = x - \frac{x^3}{6} + \frac{x^5}{120} - \frac{x^7}{5040} + \frac{x^9}{362{,}880} - \frac{x^{11}}{39{,}916{,}800} + \frac{x^{13}}{6{,}227{,}020{,}800} - \frac{x^{15}}{1{,}307{,}674{,}368{,}000}.$$

The calculator happily grinds out the sum, which we then accept as an approximation for $\sin(0.5)$. Why?

The gimmick works because that 15th-degree polynomial's graph (shown dotted below) is virtually indistinguishable from the sine wave, at least for small values of x. Indeed, the two graphs are *so* close that for inputs between $-\pi/2$ and $\pi/2$, the polynomial's output matches sine's in the first nine decimal places – precisely what the calculator displays on its screen. The calculator is a very good faker indeed!

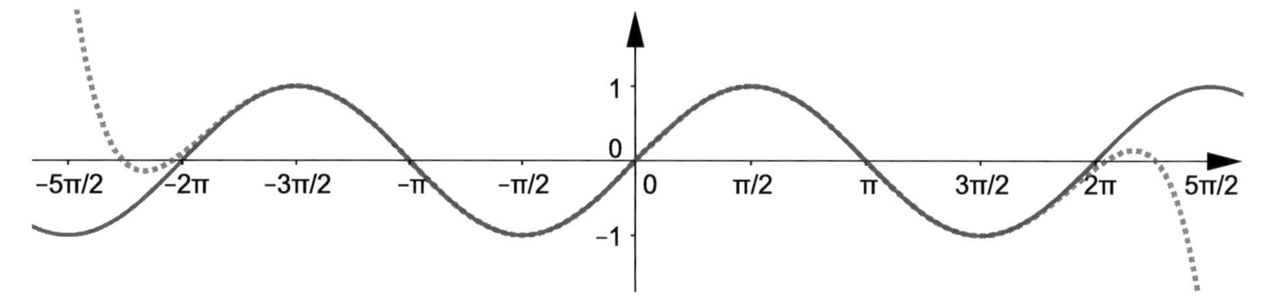

Your calculator's programmer knows how to mimic *any* well-behaved function with a polynomial – at least for inputs in the vicinity of zero.† You too will know how to do this soon.

Problem. Given a function f, find a polynomial p such that $p(x) \approx f(x)$ for all x *near* 0.

Solution. We'll begin by listing properties we'd like p to have in an infinitesimal neighborhood of 0. First, we know that our approximation will weaken as x moves away from zero, but *at* zero, the approximation should be perfect. Thus, we want p to have this property: $\boldsymbol{p(0) = f(0)}$.

Next, we'd like p and f to pass through the vertical axis with the same slope, for then the functions will not only "start" at the same value, but they'll also be changing *from* that equal value at equal rates – at least initially. Thus, we want p to have this property: $\boldsymbol{p'(0) = f'(0)}$.

Of course, the two functions' initially equal *rates of change* will eventually drift apart, leading to the separation of the functions themselves. Still, we can hold the functions together a bit longer by making our third request: equality of the *rates* at which the functions' rates of change are changing at 0. (Read that again slowly.) Thus, we'll want this property: $\boldsymbol{p''(0) = f''(0)}$.

* In particular, calculators can't understand geometry. What is π? You know that π is the circumference-to-diameter ratio in any circle. But the poor calculator (like many a poor student) only "knows" π as a string of digits someone has programmed into it.

† Actually, there's nothing special about 0. We could center the approximation at a different point, as you'll see in the exercises.

Even if these three stringent demands are met, the function f and the polynomial p will part ways eventually, but by now you've surely guessed how we can keep them close together for as long as possible: **At zero, we want p to agree with f on as many derivatives as possible.**

This derivative-matching strategy will lead us directly to the polynomial we seek. If we want p to be a polynomial of degree n, then, by definition, it must have the form

$$p(x) = c_0 + c_1 x + c_2 x^2 + c_3 x^3 + \cdots + c_n x^n. \qquad \bigstar$$

Our problem is to find the coefficients. By matching derivatives, we can find them one at a time.

Let's begin. We want $p(0) = f(0)$, but the starred equation implies that $p(0) = c_0$. Equating these two expressions for $p(0)$ reveals that the polynomial's constant term must be $\boldsymbol{c_0 = f(0)}$.

We want $p'(0) = f'(0)$. Direct computation with the starred equation shows that $p'(0) = c_1$. (Verify this.) Equating the two expressions for $p'(0)$ reveals that $\boldsymbol{c_1 = f'(0)}$.

We want $p''(0) = f''(0)$. Computation yields $p''(0) = 2c_2$, so we must have $\boldsymbol{c_2 = f''(0)/2}$.

We want $p'''(0) = f'''(0)$. Computation yields $p'''(0) = (3!)c_3$, so we have $\boldsymbol{c_3 = f'''(0)/3!}$.

Continuing this way, you'll quickly discern the pattern. For our approximating polynomial to meet our demands, its coefficient for x^k (for all k) must be the function's kth derivative evaluated at zero, divided by k factorial. Or in symbols, $\boldsymbol{c_k = f^{(k)}(0)/k!}$.*

Thus, we conclude that

> The nth-degree polynomial that best approximates $f(x)$ near zero is:
>
> $$\boldsymbol{p(x) = f(0) + f'(0)x + \frac{f''(0)}{2!}x^2 + \frac{f'''(0)}{3!}x^3 + \cdots + \frac{f^{(n)}(0)}{n!}x^n.}$$
>
> This is called a *Taylor polynomial* for the function f.

More specifically, the boxed function is called *the nth degree Taylor polynomial for f centered at* 0. Note the three modifiers: they specify the polynomial's degree, the function it approximates, and the "center" of the approximation (i.e. the point at which we play the derivative-matching game). ♦

Now we can see where our mendacious calculator's 15th-degree Taylor polynomial for sine came from. Sine's derivatives, starting with the "zeroth", cycle through four functions: $\sin x$, $\cos x$, $-\sin x$, $-\cos x$. When we evaluate these derivatives at zero, we get an endlessly repeating cycle of numbers: $0, 1, 0, -1$. What do these numbers represent? According to the boxed formula above, they are the *numerators* in sine's Taylor polynomial. Note that the two zeros in the cycle of numerators make every other term in the polynomial vanish, which is why the polynomial has only odd-power terms.

The greater the number of terms in a Taylor polynomial, the better it approximates its target function. We can see this by plotting several Taylor polynomials for sine and observing their increasingly impressive mimicry.

* To ensure that this formula works even for c_0, we'll adopt the convention that the "zeroth derivative" of a function is the function itself. You probably know already that $0!$ is *defined* to be 1. (And if you didn't know it, well, now you do.)

If we let T_k represent the kth-degree Taylor polynomial for sine (centered at zero), then, to take just a few examples, we have

$$T_1(x) = x$$

$$T_3(x) = x - \frac{x^3}{3!}$$

$$T_5(x) = x - \frac{x^3}{3!} + \frac{x^5}{5!}$$

and

$$T_7(x) = x - \frac{x^3}{3!} + \frac{x^5}{5!} - \frac{x^7}{7!}.$$

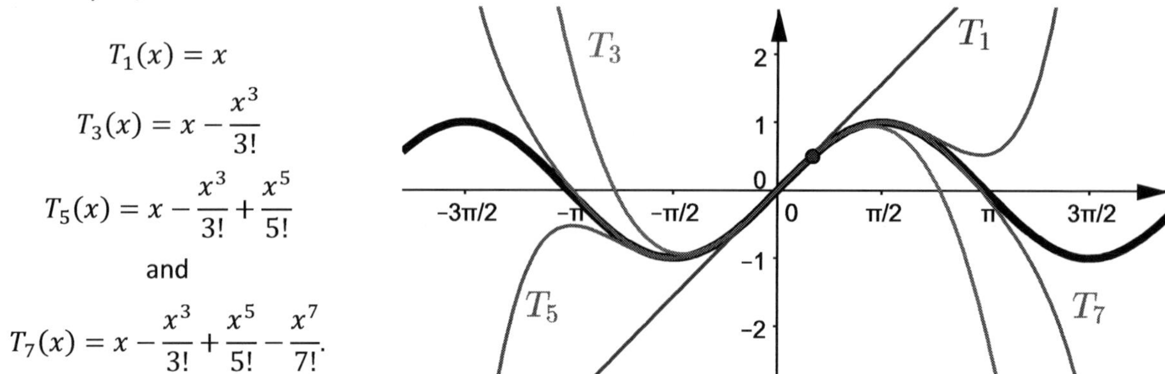

As you can see, even the first-order Taylor approximation, $\sin x \approx x$, is surprisingly decent when x is small. But note well: Except at the origin, these Taylor polynomials always differ slightly from the sine wave. Consider, for example, the point $(\pi/6, 1/2)$, which I've emphasized above. It lies on sine's graph, and *seems* to lie on the polynomials' graphs, but in fact not one of the Taylor polynomials passes through it. They merely come close to the point, and the higher the polynomial's degree, the closer the approach. Magnifying the vicinity of the point enormously, we obtain the boxed picture at right, which represents a region approximately $.0001$ units high and $.0003$ units wide. On this miniscule scale, we see not only that T_3 and T_5 miss the point, but also that T_5 approximates sine's graph much better than T_3 does. But where are the graphs of T_1 and T_7? The graph of T_1 doesn't pass through the boxed region at all, while the graph of T_7 is invisible because even on this tiny scale it is *still* indistinguishable from sine. And yet, if we magnify the region near $(\pi/6, 1/2)$ still further, the gap between T_7 and sine will become apparent.* I encourage you to see this for yourself by playing around with a graphing program; it will give you a more visceral feel for this phenomenon.

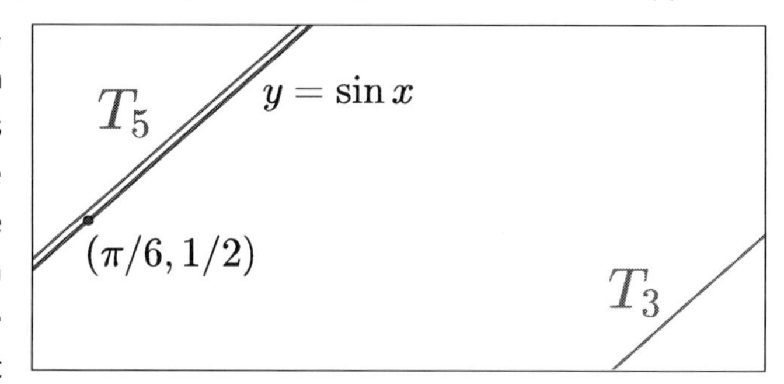

The figures above suggest that increasing a Taylor polynomial's degree expands the interval over which it accurately approximates its target function. For example, T_1's mimicry of sine looks fairly accurate over the interval $(-\pi/4,\ \pi/4)$, while T_7's looks accurate over $(-\pi,\ \pi)$. And looking back at the graph on this chapter's first page, we see that T_{15}'s approximation seems to be quite accurate over $(-2\pi,\ 2\pi)$. Still, this hint of an ever-expanding "interval of accurate approximation" is, for now, only a suggestion. We must be on guard; pictures can be deceiving, and it is dangerously easy to overgeneralize on the basis of a few examples. For some functions (including sine), we can indeed expand the interval of accurate approximation indefinitely. Other functions, however, seem to possess a security system against Taylor approximations; it's as if they have laser trip wires at certain points in their domains. When a Taylor polynomial's graph crosses such a point, it ceases to approximate the function, no matter how high its degree. Here's an example of this phenomenon.

* For instance, at $\pi/6$ itself, the values of sine and T_7 differ by about 0.000000008. A tiny difference, but still visible with enough magnification.

Example. Find and graph some Taylor polynomials (centered at zero) for $f(x) = \frac{1}{1+x^2}$.

Solution. To find the coefficients of such polynomials, we'll need to know the function's derivatives, and their values at zero. Slogging through these derivatives and cleaning up the results (note the way I've organized the results; you'll find it helpful to do the same when you compute your own Taylor polynomials in the exercises), we find that

$$f(x) = \frac{1}{1+x^2} \qquad f(0) = 1$$

$$f'(x) = \frac{-2x}{(1+x^2)^2} \qquad f'(0) = 0$$

$$f''(x) = \frac{6x^2-2}{(1+x^2)^3} \qquad f''(0) = -2$$

$$f'''(x) = \frac{-24x(x^2-1)}{(1+x^2)^4} \qquad f'''(0) = 0$$

$$f^{(4)}(x) = \frac{24(5x^4-10x^2+1)}{(1+x^2)^5} \qquad f^{(4)}(0) = 24$$

$$f^{(5)}(x) = \frac{-240x(3x^4-10x^2+3)}{(1+x^2)^6} \qquad f^{(5)}(0) = 0$$

$$f^{(6)}(x) = \frac{720(7x^6-35x^4+21x^2-1)}{(1+x^2)^7} \qquad f^{(6)}(0) = -720.$$

Putting these values into our general Taylor polynomial formula, we find that

$$T_6(x) = 1 - x^2 + x^4 - x^6.$$

One can prove that this pattern holds indefinitely, so that, for example,

$$T_8(x) = 1 - x^2 + x^4 - x^6 + x^8.$$

The graphs of the original function and a few Taylor polynomials (centered at 0) are shown at right. ♦

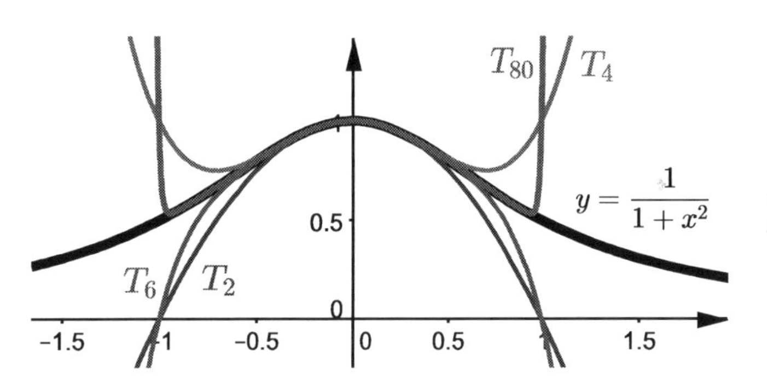

As the graph of T_{80} above suggests, the function $1/(1+x^2)$ is approximated by its Taylor polynomials (centered at zero) only over $(-1, 1)$. To take up our earlier analogy, it's as if this function has laser trip wires installed at $x = -1$ and $x = 1$; when a Taylor polynomial passes over either point, its ability to approximate its target function is destroyed, no matter how high its degree. Sine, in contrast, has no such security system. The 80th-degree Taylor polynomial for sine, for example, hugs the sine wave closely as it rolls through *ten* full periods. There is even a sense, to be discussed later in this chapter, in which sine's Taylor polynomial of *infinite* degree (a Taylor *series*) matches the sine function over the whole real line.

Why this dramatic difference in behavior between sine and $1/(1+x^2)$? To answer this question, we'll need to dig deeper into all things Taylor. But before we resume the excavation, do these exercises.

Exercises.

1. For each of the following functions, find the 7th-degree Taylor polynomial centered at zero. The Taylor expressions of these four functions are particularly simple, and particularly important, so commit them to memory right away.

a) $\sin x$ b) $\cos x$ c) e^x d) $1/(1-x)$

2. Find the 5th-degree Taylor polynomial approximation for $\tan x$ centered at zero. The moral of this exercise: The coefficients in a function's Taylor approximation need not follow a simple pattern.*

3. If p is a polynomial of degree n, what is its nth-degree Taylor polynomial centered at zero? Explain why this makes intuitive sense.

4. For any reasonably nice function f, there is an nth-degree Taylor polynomial for it *centered at* a (instead of at 0) for any a in its domain. This "re-centering" is a simple matter of anchoring the Taylor polynomial to f at $(a, f(a))$ instead of at $(0, f(0))$. Done properly (by matching all derivatives **at** $\boldsymbol{a}$), this ensures that our Taylor polynomial approximates f well when x is near a, but of course, the approximation's quality lessens as the distance between x and a increases. The figure below, for example, shows the sine wave and several Taylor polynomials for it centered at $\pi/2$.

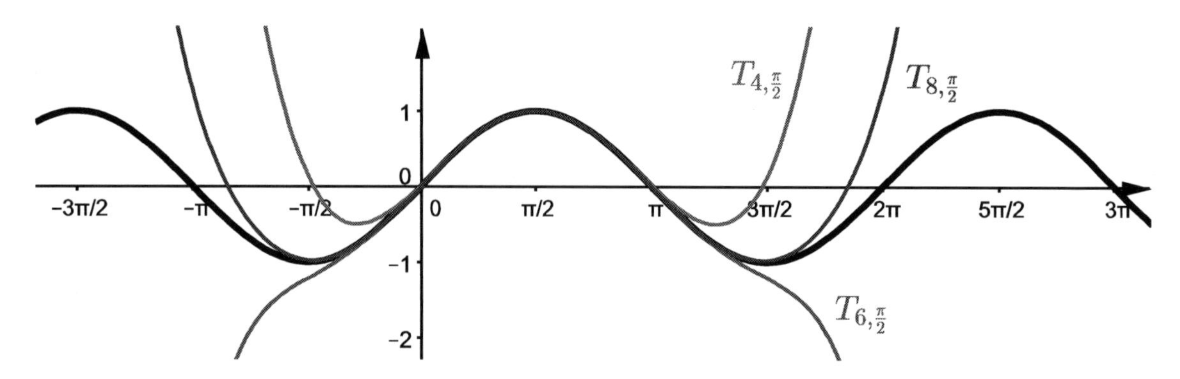

Since the quality of the approximation at any given point x depends on x's distance from a, it is convenient to express the Taylor polynomial's terms as powers of $(x-a)$ instead of as powers of x. By playing the derivative-matching game, one can show that the nth-degree Taylor polynomial for f centered at a must be:

$$T_{n,a}(x) = f(a) + f'(a)(x-a) + \frac{f''(a)}{2!}(x-a)^2 + \frac{f'''(a)}{3!}(x-a)^3 + \cdots + \frac{f^{(n)}(a)}{n!}(x-a)^n,$$

which is a formula you should certainly know. Some problems:

a) Convince yourself that $T_{n,a}$ does in fact match f on its first n derivatives at a.

b) Find the 6th-degree Taylor polynomial for sine centered at $\pi/2$.

c) Find the 3rd-degree Taylor polynomial for e^x centered at $\ln 2$.

5. In Exercise 1, you saw that sine, an *odd* function, has only odd-degree terms in its Taylor polynomial, while cosine, an *even* function, has only even-degree terms in its Taylor polynomial. This was no coincidence.

a) All even functions have odd derivatives. Explain geometrically why this is true.

b) All odd functions have even derivatives. Explain geometrically why this is true.

c) Every odd function that is defined at 0 passes through the origin. Explain why. [*Hint*: $0 = -0$.]

d) Using this problem's previous parts, explain why the Taylor polynomials (centered at zero) of even functions can contain only even-degree terms, while those of odd functions can contain only odd-degree terms.

---

* And yet, there *is* a subtle pattern to the coefficients of tangent's Taylor expansion. It involves the so-called *Bernoulli numbers*, which are important in various fields of advanced mathematics.

The Error in a Taylor Approximation

Approximation always entails error. Provided we know that the error is insignificant, we often don't mind. When we use a Taylor polynomial to approximate a function's value at a point x, we can't say exactly how much error has crept in, but a handy "error bound theorem" for Taylor polynomials can often assure us that the error is *at most* such-and-such. For example, substituting 0.3 into the 5th-degree Taylor polynomial for sine yields $\sin(0.3) \approx 0.29552025$. How good is this approximation? As we'll soon see, our error bound theorem can guarantee that it differs from the true value by *less than* 0.000002.*

I'll prove the Error Bound Theorem after I've discussed how to use it. Before I can even state the theorem, though, I'll need to introduce some terminology and notation.

If $T_{n,a}$ is the nth-degree Taylor approximation to a function f (centered at a), then for any x in f's domain, we define the **error** of the Taylor approximation at x to be the quantity

$$E_n(x) = f(x) - T_{n,a}(x).$$

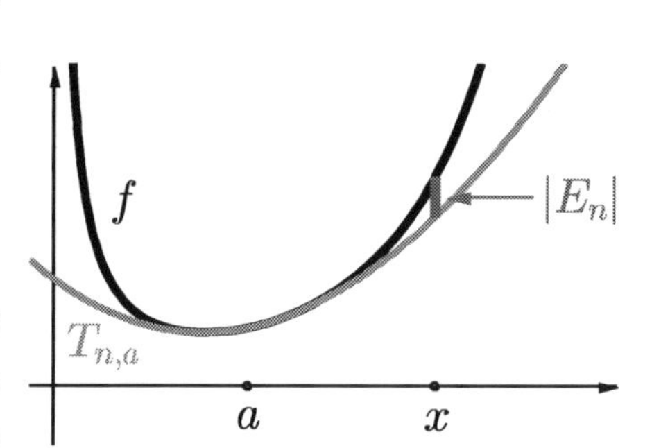

Our main concern will be with the error's *magnitude*, $|E_n(x)|$; its sign is largely irrelevant for our purposes. Observe on the graph at right that $|E_n(x)|$ is a genuine function of x. At a, its value is 0; as we move further from a, the error's magnitude increases.

> **The Error Bound Theorem.** Suppose we are approximating f with $T_{n,a}$ (its Taylor polynomial of degree n centered at a) at some point b. If we can find a number M such that $|f^{(n+1)}(x)| \leq M$ for all x between a and b, then the error's magnitude will be bounded as follows:
>
> $$|E_n(b)| \leq \frac{M}{(n+1)!}|b-a|^{n+1}.$$

Yes, that is a complicated statement. But don't miss the key idea: To find an upper bound for the error, we must first find an upper bound M for the $(n+1)$st derivative of the function we are approximating.

For our first example, let's go back and justify the claim I made earlier about $\sin(0.3)$.

Example 1. Estimate $\sin(0.3)$ with the 5th-degree Taylor polynomial for sine, and find an upper bound on the estimate's error.

Solution. Here, $f(x) = \sin x$. Using $T_5(x)$, we approximate its value at $b = 0.3$, and find that

$$\sin(0.3) \approx T_5(0.3) = 0.3 - \frac{(0.3)^3}{3!} + \frac{(0.3)^5}{5!} = 0.29552025.$$

* You might well ask, "Who needs the error bound theorem? I'll just check the approximation's value against my calculator's." Ah, but how do you know that *its* values are accurate? "Because if my calculator's digits weren't accurate, the company that makes it would go out of business," you might reply – and you'd be entirely correct. Faith in capitalism is a fine thing, but you shouldn't let it limit your understanding of mathematics. "Any idiot can know," said Einstein. "The point is to *understand*."

To use the error bound theorem, we'll need an upper bound M on all the values of $|f^{(6)}(x)|$ as x runs from 0 to 0.3. Such a bound is easily found: Since $f^{(6)}(x) = -\sin x$, we know $|f^{(6)}(x)| \leq 1$ for *all* x (including, of course, all x between 0 and 0.3). Hence, we may take $M = 1$ in this case.

Hence, by the error bound theorem, we have

$$|E_5(0.3)| \leq \frac{1}{6!}|0.3 - 0|^6 = 0.0000010125.$$

Stated more conservatively (for the sake of tidiness), **the error is no larger than than $\mathbf{0.000002}$.**

Since our estimate (0.29552025) is now known to be within 0.000002 of the true value, we know that $\sin(0.3)$ must lie between 0.29551825 and 0.29552225. We have therefore definitively established (without having needed to appeal to a calculator's authority at any stage) the first four decimal places of $\sin(0.3)$. ♦

In Chapter 2, you investigated e's numerical value as follows: First, you proved that $e \approx (1 + (1/n))^n$ when n is large; then you substituted in a large value of n.* With better tools at our disposal, we can now approximate e another way: Evaluate a Taylor polynomial for e^x at $x = 1$. This new method has two advantages. First, it's conceptually simpler. Second, it lets us quantify our approximation's accuracy via the error bound theorem.

Example 2. Using a 10th-degree Taylor polynomial for e^x, find bounds for e's numerical value. (All you should assume about e at the outset is that $2 < e < 3$.)

Solution. Let $f(x) = e^x$, and let $T_{10}(x)$ be its 10th-degree Taylor polynomial centered at 0. Then

$$e = e^1 \approx T_{10}(1) = 1 + 1 + \frac{1}{2!} + \frac{1}{3!} + \frac{1}{4!} + \frac{1}{5!} + \frac{1}{6!} + \frac{1}{7!} + \frac{1}{8!} + \frac{1}{9!} + \frac{1}{10!} = 2.718281801 \ldots$$

To bound this estimate's error, we'll need an upper bound M for $|f^{(11)}(x)|$ as x runs from 0 to 1. Since $f^{(11)}(x) = e^x$, it is clear that as x runs from 0 to 1, the value of $|f^{(11)}(x)|$ runs from a minimum of $|e^0| = 1$ to a *maximum* of $|e^1| = e$... which we know is less than 3. Consequently, we may use $M = 3$ as our upper bound for $|f^{(11)}(x)|$ for all x between 0 and 1.

Hence, by the error bound theorem, we have

$$|E_{10}(1)| \leq \frac{3}{11!}|1 - 0|^{11} < 0.0000001.$$

Since e's true value lies within 0.0000001 of our approximation, we may conclude that

$$2.7182817 < e < 2.71828191.$$

In particular, this demonstrates that e's decimal expansion must begin 2.718281. ♦

Having explained *how* the error bound theorem works, I'll turn next to *why* it works. The proof I'll present looks terribly complicated, but isn't really. It has, in fact, a strange beauty for the connoisseur, but tyros should feel free to skip it, dig into the exercises, and return to the proof on a rainy day.

* Ch.2, exercise 30. Being older and wiser, we'd now describe our work in that problem as showing that $\lim_{n\to\infty}(1 + 1/n)^n = e$.

The proof involves *iterated integrals*, such as $\int_0^1 \left(\int_0^x t^2 dt \right) dx$, which are rife in multivariable calculus. Such integrals are nothing new: The integral in parentheses is itself a function of x (namely, $x^3/3$), so we can, of course, integrate it as x runs from 0 to 1. In practice, we omit the parentheses to avoid clutter:

$$\int_0^1 \int_0^x t^2 \, dt \, dx = \int_0^1 \frac{x^3}{3} dx = \frac{1}{12}.$$

If you understand that much, you can grasp the use of iterated integrals in the following proof.

Error Bound Theorem. Suppose we are approximating f with $T_{n,a}$ (its Taylor polynomial of degree n centered at a) at some point b. If we can find a number M such that $|f^{(n+1)}(x)| \leq M$ for all x between a and b, then the error's magnitude will be bounded as follows:

$$|E_n(b)| \leq \frac{M}{(n+1)!} |b-a|^{n+1}.$$

Proof. By the Fundamental Theorem of Calculus, $\int_a^x f'(t)dt = f(x) - f(a)$, so we can write

$$\begin{aligned} f(x) &= f(a) + \textstyle\int_a^x f'(t)dt \\ &= f(a) + \textstyle\int_a^x \left[f'(a) + \int_a^t f''(u)\, du \right] dt && \text{(by the same sort of FTC trickery)} \\ &= f(a) + \textstyle\int_a^x f'(a)dt + \int_a^x \int_a^t f''(u) du \, dt && \text{(by a linearity property of integrals)} \\ &= \boldsymbol{f(a) + f'(a)(x-a)} + \textstyle\int_a^x \int_a^t f''(u) du \, dt. \end{aligned}$$

This is starting to look familiar! We've established the following: $f(x) = \boldsymbol{T_{1,a}(x)} +$ (extra stuff). That "stuff" is our error term, $E_1(x)$. By repeating these FTC shenanigans, we can generate a Taylor polynomial of any degree we wish. Let's, as a demonstration, carry it out one step further:

$$\begin{aligned} f(x) &= f(a) + f'(a)(x-a) + \textstyle\int_a^x \int_a^t f''(u) du \, dt && \text{(from above)} \\ &= f(a) + f'(a)(x-a) + \textstyle\int_a^x \int_a^t \left[f''(a) + \int_a^u f'''(v)\, dv \right] du \, dt && \text{(by the FTC)} \\ &= f(a) + f'(a)(x-a) + \textstyle\int_a^x \int_a^t f''(a) du \, dt + \int_a^x \int_a^t \int_a^u f'''(v) dv \, du \, dt && \text{(linearity)} \\ &= f(a) + f'(a)(x-a) + \tfrac{f''(a)}{2}(x-a)^2 + \textstyle\int_a^x \int_a^t \int_a^u f'''(v) dv \, du \, dt. \\ &= \boldsymbol{T_{2,a}(x)} + \textstyle\int_a^x \int_a^t \int_a^u f'''(v) dv \, du \, dt. \end{aligned}$$

If we repeat this process until we've produced the nth-degree Taylor polynomial, we end up with

$$f(x) = \boldsymbol{T_{n,a}(x)} + \int_a^x \int_a^t \int_a^u \cdots \int_a^\xi \int_a^\phi \int_a^\psi f^{(n+1)}(\omega) \, d\omega \, d\psi \, d\phi d\xi \cdots du \, dt.$$

Since the iterated integral equals $f(x) - T_{n,a}(x)$, it equals $E_n(x)$. Hence, letting $x = b$, we have

$$E_n(b) = \int_a^b \int_a^t \int_a^u \cdots \int_a^\xi \int_a^\phi \int_a^\psi f^{(n+1)}(\omega) \, d\omega \, d\psi \, d\phi d\xi \cdots du \, dt.$$

This expression for $E_n(b)$ marks the end of the proof's first half. In the second, we'll bound $E_n(b)$.

We're now assuming that we've found a number M such that the magnitude of $f^{(n+1)}$ is always less than or equal to M. In particular, for any ω in the interval between a and b, we know that

$$-M \leq f^{(n+1)}(\omega) \leq M.$$

Integrating this inequality's terms (repeatedly, building the center term up to $E_n(b)$), we obtain

$$\int_a^b \int_a^t \int_a^u \cdots \int_a^\xi \int_a^\phi \int_a^\psi -M \, d\omega \, d\psi \, d\phi d\xi \cdots du \, dt \leq E_n(b)$$

$$\leq \int_a^b \int_a^t \int_a^u \cdots \int_a^\xi \int_a^\phi \int_a^\psi M \, d\omega \, d\psi \, d\phi d\xi \cdots du \, dt .^*$$

Working out the iterated integrals, which requires some dedicated bookkeeping, reduces this to

$$\frac{-M}{(n+1)!}(b-a)^{n+1} \leq E_n(b) \leq \frac{M}{(n+1)!}(b-a)^{n+1},$$

from which follows the truth of the statement we are trying to prove:

$$|E_n(b)| \leq \frac{M}{(n+1)!}|b-a|^{n+1}. \qquad \blacksquare$$

Exercises.

6. Using a 5th-degree Taylor polynomial for e^x, find upper and lower bounds for the numerical value of $\sqrt{e}$. (All you should assume about e at the outset is that $2 < e < 3$.)

7. To improve the estimate in the previous problem, which would make a bigger difference: increasing the Taylor polynomial's degree, or assuming at the outset that we knew that $2.718 < e < 2.719$? Play with this until you can explain intuitively why your answer is correct.

8. In example 1 above, we estimated the value of $\sin(0.3)$ with a 5th-degree Taylor polynomial for sine and deduced that our estimate was within $.000002$ of the correct value. In this case, we can improve our bound by noticing that our approximating polynomial, which we called T_5, is in fact T_6, the 6th-degree Taylor polynomial for sine. Explain why this is so, and then use this observation to find an improved error bound for the approximation.

9. Exponential functions grow faster than power functions (see Ch. π, exercise 10), but factorials grow faster still. It's easy to see why: By increasing the argument in, say, the exponential function 50^n, we keep producing more factors of 50; by increasing it in the factorial function $n!$, we produce more factors, *which themselves are growing*. The two functions 50^n and $n!$ both have n factors; in the exponential function, each of the factors is 50, but in the factorial function, when n is sufficiently large, most of the factors will be *far larger* than 50. This will ensure that the factorial function eventually overtakes the exponential function in their race toward infinity.

This being so, evaluate the following limits:

a) $\lim_{n\to\infty} \frac{10^n}{n!}$ b) $\lim_{n\to\infty} \frac{999^n}{(n+1)!}$ c) $\lim_{n\to\infty} \frac{n!}{100{,}000^n}$ d) $\lim_{n\to\infty} \frac{1001^{n+1}}{(n+1)!}$

---

* Fussy note: If $b < a$, we must alter the proof slightly: The integrals (from a to whatever) will all run "backwards" (right to left). Doing an integral backwards is equivalent to multiplying its value by -1, so if $b < a$ *and* there are an odd number of integrals (i.e. if n is even), each term in the asterisked inequality will be $(-1)^{n+1} = -1$ times what's written above. Multiplying through by -1 eliminates these negatives, but *reverses* the inequalities' direction; they will remain reversed in the next step, yielding $\frac{-M}{(n+1)!}(b-a)^{n+1} \geq E_n(b) \geq \frac{M}{(n+1)!}(b-a)^{n+1}$, from which it follows that $|E_n(b)| \leq \frac{M}{(n+1)!}|b-a|^{n+1}$, as claimed.

10. In the previous section, we saw that increasing the degree of sine's Taylor polynomial extends the interval over which its graph closely hugs the sine wave. Can we extend this "interval of accurate approximation" indefinitely by taking higher and higher degree Taylor polynomials? For example, can we jack up the degree so high that our Taylor polynomial (centered at zero) can approximate $\sin(1000)$ to within 0.001 of its true value? Let us see.

a) Show that $\lim_{n\to\infty} |E_n(1000)| = 0$, and explain how this settles our $\sin(1000)$ question. [*Hint: Exercise 9d.*]

b) Now settle our question about *indefinitely* extending the interval of accurate approximation by showing that $\lim_{n\to\infty} |E_n(x)| = 0$ for *every* real number x.

11. Suppose we want to approximate square roots near 1, such as $\sqrt{1.04}$.

a) Explain why using a Taylor polynomial for $f(x) = \sqrt{x}$ centered at zero can't help us.

b) Find the first-degree Taylor polynomial for $f(x) = \sqrt{x}$ centered at 1. Use it to approximate $\sqrt{1.04}$, and then use the Error Bound Theorem to show that your approximation is within $.0002$ of the true value. [Hint: To find an appropriate "M" to use in the Error Bound Theorem, sketch a graph of $y = f''(x)$.]

c) Despite part (a), there is in fact a sneaky way to use polynomials centered at zero to approximate square roots. First, we shift the graph of $f(x) = \sqrt{x}$ to the left by one unit, obtaining the graph of $g(x) = \sqrt{x+1}$. We can then use a Taylor polynomial of this new function *centered at zero* to approximate square roots. Draw a picture to convince yourself that you understand what I'm describing. Find the first-degree Taylor polynomial for g centered at zero, and use it to approximate $\sqrt{1.028}$. [Note that $\sqrt{1.028}$ is not $g(1.028)$, but rather $g(.028)$.]

d) Note that the simple polynomial you found in part (c) gives a very rapid method for making quite accurate approximations to square roots near 1 in one's head. (e.g. $\sqrt{1.044} \approx 1.022$, and $\sqrt{0.96} \approx 0.98$.)

12. The eternal refrain of the pragmatic (or short-sighted) student: "When am I going to use this outside of class?" The eternal answer: You never know. The following story from *The Penguin Book of Curious and Interesting Mathematics* by David Wells (a fine collection of anecdotes about mathematicians) is apropos.*

> *During the Russian revolution, the mathematical physicist Igor Tamm was seized by anti-communist vigilantes at a village near Odessa where he had gone to barter for food. They suspected he was an anti-Ukrainian communist agitator and dragged him off to their leader. Asked what he did for a living he said that he was a mathematician. The skeptical gang-leader began to finger the bullets and grenades slung around his neck. "All right", he said, "calculate the error when the Taylor series approximation of a function is truncated after n terms. Do this and you will go free; fail and you will be shot". Tamm slowly calculated the answer in the dust with his quivering finger. When he had finished the bandit cast his eye over the answer and waved him on his way. Tamm won the 1958 Nobel Prize for physics but he never did discover the identity of the unusual bandit leader.*

Now go back and study the error bound theorem's proof. It could save your life.

* Here, Wells is quoting John Barrow's article, "It's All Platonic Pi in the Sky", *Times Educational Supplement*, 11 May 1993.

Taylor Series

Each Taylor polynomial approximates a function, and by increasing the polynomial's degree, we improve the approximation, at least for values of x within some ill-defined "interval of accurate approximation". (We'll be able to define it carefully by the chapter's end.) Having imbibed the spirit of calculus for some time now, you will already have guessed where this steadily improving approximation is heading. If we push it to extremes by taking a Taylor polynomial of *infinite* degree, the approximation will become *exact.*[*] That is, the infinite-degree polynomial will no longer merely approximate the function; it will actually *be* the function, albeit re-expressed in a radically new way. Such an infinite-degree Taylor polynomial is called a **Taylor series** for the function. Here, for example, is the Taylor series for e^x (centered at 0):

$$e^x = 1 + x + \frac{x^2}{2!} + \frac{x^3}{3!} + \cdots .$$

The ellipsis at the end indicates the summation continues in the same pattern, *without end*. Much of the rest of this chapter will be devoted to making sense of this jarring idea.

Taylor series are functions. To understand them, we must understand what it means to evaluate them. For instance, we can, in theory, find the numerical value of e by letting $x = 1$ in the Taylor series for e^x above. Doing so yields this mysterious equation:

$$e = 1 + 1 + \frac{1}{2!} + \frac{1}{3!} + \cdots .$$

The right-hand side is an **infinite series**, a sum of infinitely many real numbers.[†] Some infinite series, such as this one, add up to a finite value. These are called *convergent* series. In contrast, *divergent* series don't add up to a finite value. (For example, $1 + 1 + 1 + \cdots$ is clearly divergent.) At the end of the day, convergent series are numbers. Divergent series are not. Since functions (and hence Taylor series) must turn numbers into numbers, we need to be able to guarantee that when we evaluate a Taylor series at a given number, the infinite series produced thereby is convergent. Thus, in the next several sections, we'll discuss various tests for convergence of infinite series. Then we shall return to Taylor series.

Exercises.

13. Taylor series centered at 0 are often called *Maclaurin series*. Find Maclaurin series for the following functions.

a) e^x b) $1/(1-x)$ c) $\sin x$ d) $\cos x$ e) $\ln(1+x)$ f) $(1+x)^\alpha$, where α is a constant.

[*By now, you should be able to retrieve the first four from memory. For the last two, you'll need to crank out some derivatives. The last one, incidentally, is the famous* ***binomial series***.]

14. Rewrite the series you found in parts (a) – (e) in the previous exercise with sigma summation notation.

15. With help from a calculator, add up the first ten terms in the series for e given above. Observe how with each successive term, the running total gets closer and closer to e.

---

[*] Compare the idea of the definite integral as the limit of Riemann sums (in which an approximation to an area becomes exact as the number of approximating rectangles becomes infinite), and the idea of a derivative as the limit of a difference quotient (in which an approximation to a slope becomes exact as Δx become infinitesimal).

[†] Note this contrast: A definite integral is an infinite sum of *infinitesimals*. An infinite series is an infinite sum of *real numbers*.

Infinite Series in General, Geometric Series in Particular

"There you stand, lost in the infinite series of the sea, with nothing ruffled but the waves."
- Herman Melville, *Moby Dick* (Ch. 35, "The Mast Head")

Any fool knows what it means to add up 5 numbers, 10 numbers, or even 50 billion numbers. If the fool also knows about limits, he can understand what it means to sum up an infinite series, too. Here's how: Given an infinite series, we let s_n be the sum of its first n terms. (We call s_n the series' **n[th] partial sum.**) Now we take the limit of s_n as $n \to \infty$. *We define this limit (if it exists) to be the sum of the infinite series.* If the limit doesn't exist (or blows up to infinity), then we say that the series diverges.

For example, the nth partial sum of the series $3 + 3 + 3 + \cdots$ is obviously $s_n = 3n$. Since s_n blows up to infinity as n increases, the series diverges. In contrast, the nth partial sum of $0.3 + 0.03 + 0.003 + \cdots$ is $s_n = 0.333...333$, with n threes after the decimal point. Here, s_n approaches $1/3$ as $n \to \infty$, so this series converges (sums up) to $1/3$.

Our first substantial examples of infinite series will be "geometric series". A **geometric series** is one in which each term is some fixed multiple of its predecessor. For example,

$$2 + \frac{4}{5} + \frac{8}{25} + \frac{16}{125} + \cdots \qquad \bigstar$$

is a geometric series since each term is $2/5$ of its predecessor. Any geometric series is completely specified by its first term, a, and its fixed multiplier, r. Indeed, every geometric series has the form

$$a + ar + ar^2 + ar^3 + \cdots$$

for some a and r. To investigate the convergence of this general series, we'll consider its nth partial sum,

$$s_n = a + ar + ar^2 + \cdots + ar^{n-1}.$$

We can find a tidier algebraic expression for this partial sum with the following neat trick. Multiplying both sides of the preceding equation by r yields

$$rs_n = ar + ar^2 + \cdots + ar^{n-1} + ar^n.$$

If we subtract this equation from the previous one, all but two of the terms on the right cancel, leaving

$$s_n - rs_n = a - ar^n.$$

Finally, we solve for s_n to get a clean expression for the nth partial sum of the general geometric series:

$$s_n = \frac{a(1 - r^n)}{1 - r}\ .^*$$

This is already an interesting result, but we're concerned with it only because we want its limit as $n \to \infty$. If $|r| < 1$, then r^n vanishes in the limit, so $s_n \to a/(1 - r)$, the value to which the series converges. If $|r| > 1$, then r^n (and hence s_n) blows up, so the series diverges. (Ditto if $r = 1$; see the footnote.) To sum up (pun intended), we've just proved the following theorem.

---

* Unless $r = 1$, in which case $s_n = na$. (To see why, just think about what the series would look like if $r = 1$.)

Geometric series. The geometric series $a + ar + ar^2 + \cdots$ converges if and only if $|r| < 1$.
When such a geometric series converges, it converges to $\frac{a}{1-r}$.

Example 1. Having recognized earlier that the series $2 + \frac{4}{5} + \frac{8}{25} + \frac{16}{125} + \cdots$ is geometric ($r = 2/5$), we now know that it converges, since $|r| < 1$. Moreover, we know the value to which it converges:

$$2 + \frac{4}{5} + \frac{8}{25} + \frac{16}{125} + \cdots = \frac{2}{1-2/5} = \mathbf{10/3}. \quad \blacklozenge$$

Example 2. The series $\frac{1}{5} - \frac{4}{15} + \frac{16}{45} - \frac{64}{135} + \cdots$ is geometric, but *diverges*, since $r = -4/3$. ♦

Convergent geometric series are nice because we can determine the exact values to which they converge. This is rarely possible for non-geometric series. When confronted with a non-geometric series, our main concern is typically to ascertain whether it converges or diverges. If we can prove that it converges, then at least we'll know the series represents *some number*. Even if we can't determine that number's exact value, we can always approximate it with a partial sum of many terms. On the other hand, if we can show that the series diverges, then it doesn't add up to *anything*; it isn't a number at all, just an unholy mess. (Exercise 21 below shows how mistaking divergent series for numbers leads to gibberish.)

Beginning in the next section, you'll meet a variety of tests that will help you identify convergent non-geometric series, but none of these tests will help us sum the series up exactly. That being the case, you should be especially appreciative of geometric series!

Exercises.

16. Some of the following series are geometric. Some are not. Identify those that are geometric, and find the sums of the convergent geometric series.

a) $2 + \frac{2}{3} + \frac{2}{9} + \frac{2}{27} + \cdots$ b) $1 + \frac{1}{2} + \frac{1}{3} + \frac{1}{4} + \cdots$ c) $10^{-10} + 10^{-9} + 10^{-8} + 10^{-7} \cdots$

d) $1 + \frac{1}{2^2} + \frac{1}{3^2} + \frac{1}{4^2} + \cdots$ e) $\pi + 1 + \frac{1}{\pi} + \frac{1}{\pi^2} + \cdots$ f) $1 + \frac{\pi}{e} + \frac{\pi^2}{e^2} + \frac{\pi^3}{e^3} + \cdots$

17. With a calculator, find the first eight partial sums of the series $1 - (1/3!) + (1/5!) - (1/7!) + \cdots$. Does the series seem to converge? If so, to what value?

18. Find a geometric series whose first term is 5 and whose sum is 3.

19. Prove the following: If the first term of a geometric series is 5, the series *cannot* converge to 2.

20. The convergence or divergence of any given series depends upon what happens in its infinitely long "tail". The sum of the first few terms (where "few" could mean five, five trillion, or *any* finite number) is always finite. To converge, the series' terms must *eventually* get very small, very rapidly. What happens in the first "few" terms of a series, however, is irrelevant as regards its eventual convergence or divergence. With this in mind, decide whether or not the following series converge or diverge. Find the sums of those that converge.

a) $3 + 1 + \frac{2}{3} + \frac{4}{9} + \frac{8}{27} + \cdots$ b) $10 + 1 + \frac{1}{10} + \frac{1}{100} + \frac{10}{900} + \frac{100}{8100} + \frac{1000}{72900} + \cdots$

c) $1 - 2 + 3 - 4 + 2 - 1 + \frac{1}{2} - \frac{1}{4} + \frac{1}{8} - \cdots$ d) $\frac{1}{2} + \frac{1}{4} + \frac{1}{8} + \frac{1}{4} + \frac{1}{2} + 1 + 2 + 4 + 8 + \cdots$

21. Given their absurd conclusions, the following arguments obviously harbor flaws. Find them:

a) Let $s = 1 + 2 + 4 + 8 \cdots$. Then $2s = 2 + 4 + 8 + 16 + \cdots$. Consequently, $2s - s = -1$. That is, $s = -1$. Hence, we have established that $1 + 2 + 4 + 8 + \cdots = -1$.

b) Let $s = 1 - 1 + 1 - 1 + 1 - 1 + \cdots$. Then on one hand, $s = (1-1) + (1-1) + \cdots = 0 + 0 + \cdots = 0$. But on the other hand, $s = 1 - (1-1) - (1-1) + \cdots = 1 + 0 + 0 + + \cdots = 1$. Hence, $0 = 1$.

22. From exercises 1d and 13b, you know the Taylor series for $1/(1-x)$, which you should have memorized by now:

$$\frac{1}{1-x} = 1 + x + x^2 + x^3 + \cdots .$$

Observe that, for any value of x, this Taylor series is *geometric*.

a) For which values of x will the series converge?

b) The interval you found in the previous part is called the *interval of convergence* for that particular Taylor series. It turns out that every Taylor series has a specific interval of convergence. Inside the interval, a Taylor series equals the function from which it was derived: Given equal inputs, they produce equal outputs.* Outside the interval, the Taylor series diverges. For example, consider the equation above. When $x = -1$ (which is outside the interval), the equation's left-hand side is $1/2$, but the right-hand side is $1 - 1 + 1 - 1 + \cdots$, which diverges, and so isn't equal to *anything*, much less $1/2$.

Consequently, when we write down the Taylor series for $1/(1-x)$ in the future, we must do so with a bit more care, and express it like this:

$$\frac{1}{1-x} = 1 + x + x^2 + x^3 + \cdots, \quad \textbf{for } |x| < 1.$$

Later in this chapter, you'll learn that the interval of convergence of *any* given Taylor series depends in a subtle way upon properties of geometric series, even though most Taylor series are *not* geometric! Your problem: Keep your eyes open as this particular story quietly unfolds during this chapter.

c) The series $1 + 3x + 9x^2 + 27x^3 + \cdots$ converges for some values of x, and diverges for others. Find the value to which this series converges (this will be a function of x, of course), and find the *interval of convergence* for this series. Write up your final result carefully, using the same format as the equation near the end of part (b). [*Hint: You may find it helpful to recall a basic property of absolute values*: $|ab| = |a||b|$.]

d) Same story as part (c), but with this series: $1 - 4x^2 + 16x^4 - 64x^6 + \cdots$.

e) We can now answer a question raised earlier in the chapter: Why do the Taylor polynomials for $1/(1+x^2)$, in contrast to those of $\sin x$, seem only to work in a strictly limited "interval of accurate approximation"? To answer this, consider the Taylor series for $1/(1+x^2)$, which is $1 - x^2 + x^4 - x^6 + \cdots$. (You need not verify this; the derivatives involved are routine, but tedious.) For any x, this Taylor series is *geometric*. Use this observation to find the interval of convergence for the Taylor series of $1/(1+x^2)$.

23. We can understand the sums of the two geometric series

$$1/4 + 1/4^2 + 1/4^3 + \cdots = 1/3$$

and

$$1/3 + 1/3^2 + 1/3^3 + \cdots = 1/2$$

visually by pondering the figures at right. Ponder them.

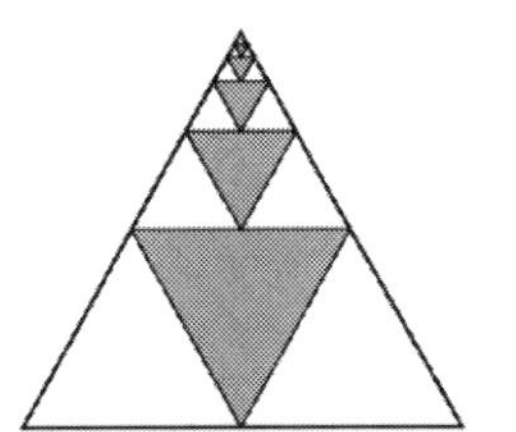

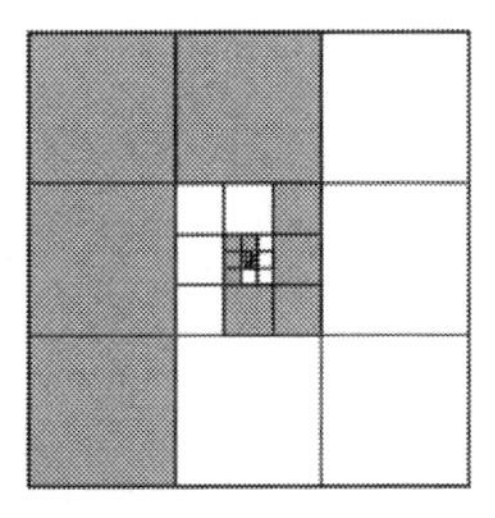

* Actually, there are some bizarre exceptions (which need not concern you in this class) called *smooth non-analytic functions*, whose Taylor series do not agree with the functions from which they were derived – even inside their intervals of convergence! Readers who go on to study real analysis will eventually meet these beasts in their native habitat, far from freshman calculus.

The Divergence Test and the Harmonic Series

"I believe that from the earth emerges a musical poetry that is, by the nature of its sources, tonal. I believe that these sources cause to exist a phonology of music, which evolves from the universal, and is known as the harmonic series."

- Leonard Bernstein, from "The Unanswered Question", a lecture series at Harvard

To say that an infinite series $a_1 + a_2 + a_3 + \cdots$ converges to a finite sum s is to say that its nth partial sum, $s_n = a_1 + a_2 + a_3 + \cdots + a_n$ gets closer and closer to s as $n \to \infty$. Obviously (think about this until it *is* obvious), this can happen only if the individual terms in the series approach zero as we go out further and further in the series. Thus, a prerequisite for convergence is that $a_n \to 0$.[*]

Restating this observation negatively lets us frame it as a simple test for divergence.

Divergence Test. For any series $a_1 + a_2 + a_3 + \cdots$, if $a_n \nrightarrow 0$, then the series diverges.

Example. Does the series $\sum \frac{n}{3n+1} = \frac{1}{4} + \frac{2}{7} + \frac{3}{10} + \cdots$ converge or diverge?[†]

Solution. It *diverges*, for its nth term is $\frac{n}{3n+1}$, which goes to $1/3$ (not 0) in the limit. ♦

Note well: The divergence test can establish only divergence, never convergence. For a series to converge, the condition $a_n \to 0$ is necessary, but not sufficient. (Compare: Knowing algebra is necessary, but not sufficient, for passing a calculus class. If Mr. X *doesn't* know algebra, he can't pass calculus. But knowing algebra isn't sufficient to guarantee that he'll pass calculus; he'll need to work hard, too.) For a series to converge, its terms must not only vanish, but vanish *quickly*. The classic example of a series whose terms vanish, yet which still diverges (because its terms don't go to zero quickly enough) is the **harmonic series,**

$$1 + \frac{1}{2} + \frac{1}{3} + \frac{1}{4} + \frac{1}{5} + \cdots.$$

Why does the harmonic series diverge? The best proof was discovered by the remarkable medieval philosopher Nicole Oresme around the year 1350. (Yes, several centuries before the invention of calculus.) Here's a retelling of his argument, which compares the harmonic series to an obviously divergent series:

Claim. The harmonic series diverges.

Proof. $1 + \frac{1}{2} + \frac{1}{3} + \frac{1}{4} + \frac{1}{5} + \frac{1}{6} + \frac{1}{7} + \frac{1}{8} + \frac{1}{9} + \frac{1}{10} + \frac{1}{11} + \frac{1}{12} + \frac{1}{13} + \frac{1}{14} + \frac{1}{15} + \frac{1}{16} + \cdots$

$$= 1 + \frac{1}{2} + \left(\frac{1}{3} + \frac{1}{4}\right) + \left(\frac{1}{5} + \frac{1}{6} + \frac{1}{7} + \frac{1}{8}\right) + \left(\frac{1}{9} + \frac{1}{10} + \frac{1}{11} + \frac{1}{12} + \frac{1}{13} + \frac{1}{14} + \frac{1}{15} + \frac{1}{16}\right) + \cdots$$

$$> 1 + \frac{1}{2} + \left(\frac{1}{4} + \frac{1}{4}\right) + \left(\frac{1}{8} + \frac{1}{8} + \frac{1}{8} + \frac{1}{8}\right) + \left(\frac{1}{16} + \frac{1}{16} + \frac{1}{16} + \frac{1}{16} + \frac{1}{16} + \frac{1}{16} + \frac{1}{16} + \frac{1}{16}\right) + \cdots$$

$$= 1 + \frac{1}{2} + \frac{1}{2} + \frac{1}{2} + \frac{1}{2} + \cdots.$$

Since this series blows up to ∞, the harmonic series (being still greater) must as well. ■

[*] Notation such as $a_n \to 0$ is just shorthand for the more typographically complex statement $\lim_{n\to\infty} a_n = 0$.

[†] Here, I'm using some additional common shorthand. The expression $\sum a_n$ is short for $\sum_{n=1}^{\infty} a_n$.

Exercises.

24. State whether each of the following series converges or diverges – or whether we haven't yet developed results that will let us decide. Justify your answers.

a) $\sum_{n=1}^{\infty}\left(\frac{2}{3}\right)^n$ b) $\sum_{n=0}^{\infty}\left(2+\left(\frac{1}{3}\right)^n\right)$ c) $\sum_{n=1}^{\infty} 1/n^2$ d) $\sum_{n=1}^{\infty} 1/n$ e) $\sum_{n=5,000}^{\infty} 1/n$

f) $\sum_{n=1}^{\infty}(1.01)^n$ g) $\sum_{n=1}^{\infty}\left(\frac{n+1}{n}\right)$ h) $\sum_{n=1}^{\infty}\left(\frac{2n+3}{9n}\right)$ i) $\sum_{n=1}^{\infty}\left(\frac{2n+3}{9n^3}\right)$ j) $2-2+2-2+\cdots$

25. Oresme's proof shows that the harmonic series blows up to infinity. That is, by adding up enough terms of the harmonic series, the running total will eventually surpass any value that we care to specify. For example, the running total passes 2 after we've summed the first four terms: $1+1/2+1/3+1/4 \approx 2.08$. After summing the first eleven terms of the series, the running total exceeds 3.

a) With a calculator, determine how many terms of the harmonic series must be summed up to surpass 4.

b) With a computer, determine how many terms must be summed up to surpass 5, 6, and 10.
[*Hint: Using Wolfram Alpha, try commands such as* "sum$(1/n)$, $n = 1$ to 40".]

c) Still using a computer, approximate (to the nearest whole number) the sum of the harmonic series' first billion terms, first trillion terms, and first 10^{82} terms. (This last is an estimate of the number of atoms in the universe.)

d) Remember the moral of the story: The harmonic series blows up to infinity, but takes its time getting there.

e) Ponder the following: In part (c), your computer just approximated a sum of 10^{82} terms. It did *not* do this by rapidly summing those terms. To be perfectly clear on this point, let us suppose you have a computer that can add a trillion trillion trillion trillion trillion trillion terms per second. (No computer can operate at a speed even close to this, and it is probably fair to assume that no computer ever will, but let us pretend.) How long would it take this fantasy computer to add up the first 10^{82} terms of the harmonic series?

26. Decide whether the following argument is a valid proof that the harmonic series diverges. If it doesn't, then what's wrong with it? If it does, then which of the two proofs that you've seen do you prefer?

Suppose the harmonic series converges to some real number S. In that case, we have

$$\begin{aligned} S &= 1+\frac{1}{2}+\frac{1}{3}+\frac{1}{4}+\frac{1}{5}+\frac{1}{6}+\frac{1}{7}+\frac{1}{8}+\cdots \\ &= \left(1+\frac{1}{2}\right)+\left(\frac{1}{3}+\frac{1}{4}\right)+\left(\frac{1}{5}+\frac{1}{6}\right)+\left(\frac{1}{7}+\frac{1}{8}\right)+\cdots \\ &> \left(\frac{1}{2}+\frac{1}{2}\right)+\left(\frac{1}{4}+\frac{1}{4}\right)+\left(\frac{1}{6}+\frac{1}{6}\right)+\left(\frac{1}{8}+\frac{1}{8}\right)+\cdots \\ &= 1+\frac{1}{2}+\frac{1}{3}+\frac{1}{4}+\cdots \\ &= S. \end{aligned}$$

*We've just shown that **if** the harmonic series converges to a real number, then this number is larger than itself. Since this is impossible, it follows that the harmonic series must not converge. Hence, it diverges, as claimed.*

27. As its name suggests, the harmonic series occurs in music theory. Learn about this.

28. Explain *why* the figure at right is a "picture proof" of the fact that

$$\sum_{n=0}^{\infty}\frac{3}{4}\left(\frac{1}{4}\right)^n = 1.$$

Then explain why this series converges to 1 by more prosaic means.

29. (Linearity Properties) Explain intuitively why the following handy properties hold:

a) $\sum ca_n = c\sum a_n$, for any constant. b) If $\sum a_n$ and $\sum b_n$ converge, then $\sum(a_n+b_n) = \sum a_n + \sum b_n$.

Comparison Tests and P-Series

"Make my funk the P-funk. I wants to get funked up."
- George Clinton

Comparison tests can help us demonstrate that a given series of *positive* terms converges (or diverges).* To prove that $\sum a_n$ converges with a comparison test, we first identify a *convergent* series $\sum c_n$, then verify that the terms of $\sum a_n$ eventually "stay below" (i.e. are always less than or equal to) those of $\sum c_n$.

To see why this comparison test works, consider the figure at right, and recall that a series converges when its terms vanish very rapidly. The solid dots' heights represent the terms of the known convergent series $\sum c_n$. Since $\sum c_n$ converges, its terms must vanish very rapidly. If, as in the figure, the terms of $\sum a_n$ stay consistently below those of the convergent series, then they vanish *even faster*. Hence, $\sum a_n$ must also converge.

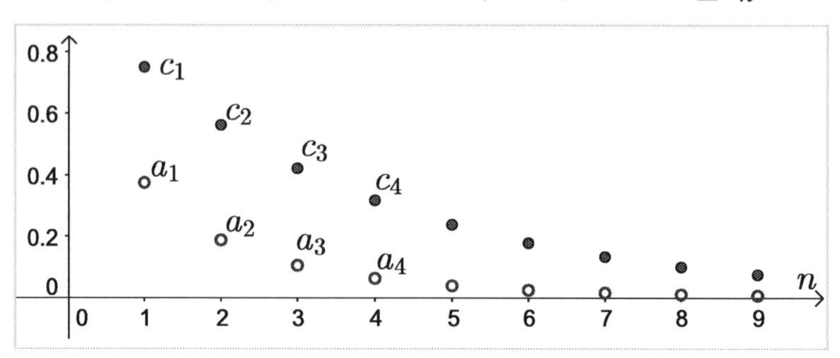

Example 1. Test the series $1+\frac{3}{8}+\frac{9}{48}+\frac{27}{256}+\cdots+\frac{3^n}{(n+1)\,4^n}+\cdots$ for convergence.

Solution. This series isn't geometric, and the divergence test doesn't apply, but a comparison test might work. Since the general term of this series resembles that of the geometric series $\sum(3/4)^n$, comparing these two series might prove fruitful. The geometric series converges, so the only question is whether the terms in the given series stay below it. If we consider the two series side by side, we see that this is indeed the case:

$$\overbrace{1+\frac{3}{8}+\frac{9}{48}+\frac{27}{256}+\cdots+\frac{3^n}{(n+1)\,4^n}+\cdots}^{\textit{The series we are testing}} \quad \text{vs.} \quad \overbrace{1+\frac{3}{4}+\frac{9}{16}+\frac{27}{64}+\cdots+\frac{3^n}{4^n}+\cdots}^{\textit{The convergent geometric series}}$$

Starting at the second term, those in the left series are always less than their mates in the right series. This pattern will hold forever, since the general term of the left series is the right series' general term *divided by a whole number*. Hence, we may conclude that the left series converges. ♦

To use the comparison test, you need a stock of well-understood series with which to make comparisons. You're already familiar with geometric series. To expand your repertoire, we'll now turn to the p-series, which boast a similarly simple convergence criterion.

If you raise all terms of the harmonic series to some fixed power p, the result is called a ***p*-series**. For example, the series $1+1/4+1/9+1/16+\cdots$ is a p-series with $p=2$. Clearly, if $p>1$, the terms of the p-series stay below those of the harmonic series... which diverges, so we can't use the comparison

* Actually, comparison tests can be used on series containing finitely many negative terms, since any such series has a *last* negative term, after which it settles down into an infinitely long tail of strictly positive terms. And as you learned in exercise 20, the convergence or divergence of a series is determined solely in the infinite tail.

test to conclude anything from this, yet we might still wonder if a large enough value of p could shrink the harmonic series' terms down far enough to force convergence. If so, how large would p have to be?

Interestingly, *any* value of p above 1 will force convergence. Indeed, the two claims below show that the harmonic series itself is the boundary between convergent p-series and divergent p-series.

Claim 1. The p-series $\sum_{n=1}^{\infty} 1/n^p$ converges whenever $p > 1$.

Proof. We will establish its convergence by comparing the p-series to an improper integral. Ponder the figure at right until you are convinced that

$$\sum_{n=2}^{\infty} 1/n^p < \int_1^{\infty} \frac{1}{x^p} dx.$$

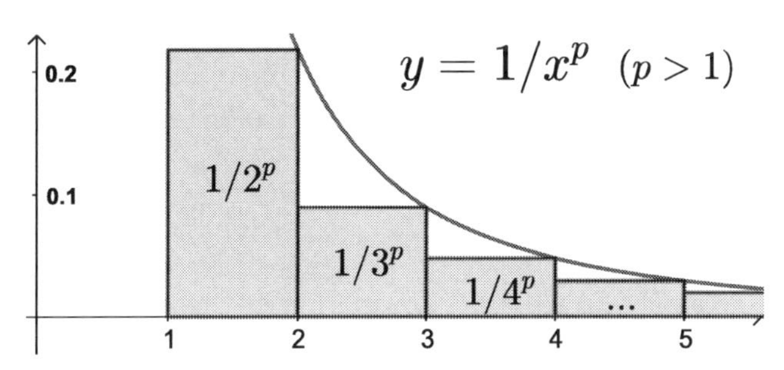

Then, by a chain of easy computations,

$$\sum_{n=2}^{\infty} 1/n^p < \int_1^{\infty} \frac{1}{x^p} dx = \int_1^{\infty} x^{-p} dx = \left[\frac{x^{1-p}}{1-p}\right]_1^{\infty} = \frac{1}{1-p}\left[\frac{1}{x^{p-1}}\right]_1^{\infty} = \frac{1}{1-p}[0-1] = \frac{1}{p-1}.$$

The sum on the left-hand side, being less than the real number $1/(p-1)$, is necessarily finite. Adding 1 to it preserves its finitude and turns it into the full p-series. (Its first term was missing.) Hence, we conclude that for *all* $p > 1$, the p-series converges to some finite value, as claimed. ■

We've seen the comparison test for convergence. Let us now consider the comparison test for *divergence*. Proving that a series of positive terms $\sum a_n$ diverges by means of a comparison test involves two steps: First, identify a series $\sum d_n$ that is *known to diverge*. Then, if the terms of $\sum a_n$ *eventually "stay above"* (i.e. are consistently greater than or equal to) the corresponding terms of the divergent series, we can conclude that $\sum a_n$ must diverge, too.

Once again, the logic behind the test is best understood by means of a picture. In the figure, the heights of the solid dots represent the terms of the known divergent series, $\sum d_n$. Since it diverges, its terms do not vanish rapidly enough for convergence to occur. That being so, it's obvious that any other series whose terms stay consistently *above* those of the divergent series must also diverge; its terms, after all, vanish *even slower* than those in the divergent series $\sum d_n$.

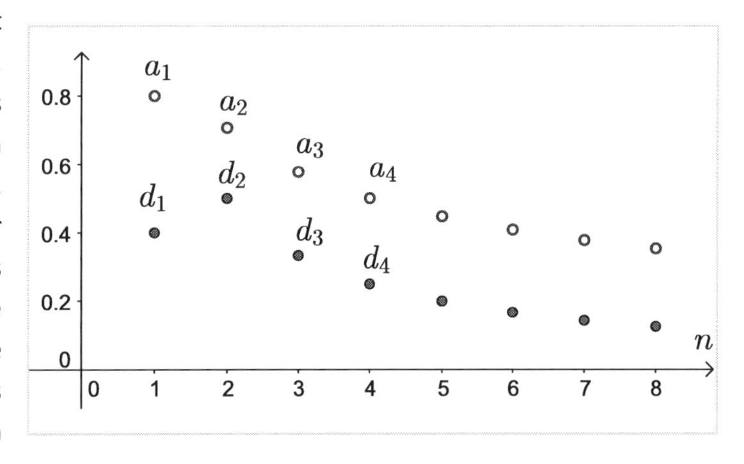

We can use the comparison test for divergence to prove that when $p < 1$, the p-series diverges.

Claim 2. The p-series diverges whenever $p < 1$.

Proof. For whole numbers n, the condition $p < 1$ implies that $n^p \leq n$, and hence that $1/n^p \geq 1/n$. Consequently, $\sum 1/n^p$ diverges by comparison with the harmonic series, $\sum 1/n$. ∎

The two comparison tests we've considered so far – the *direct* comparison tests – are conceptually simple, but somewhat awkward in practice. Juggling inequalities is rarely fun, so most people bristle when faced with the prospect of verifying that an inequality holds throughout the infinitely long tails of two series. Fortunately, in place of direct comparison, we can often use the so-called *limit* comparison test. The logic that justifies the limit comparison test is decidedly trickier than that of a direct comparison test, but the limit comparison test is often easier to use in practice, since it involves limits instead of inequalities.

The Limit Comparison Test. To test a series of positive terms $\sum a_n$ for convergence, find a related series of positive terms $\sum b_n$ whose status (convergent or divergent) you already know. Take the limit of the ratio a_n/b_n as $n \to \infty$. If this limit is a nonzero real number L, then the two series stand or fall together. That is, both converge or both diverge.

Proof. If $(a_n/b_n) \to L$, we effectively have $a_n = Lb_n$ in the series' tails, where their convergence or divergence is determined. Hence, $\sum a_n$ stands or falls with $\sum Lb_n$. Since $\sum Lb_n$ stands or falls with $\sum b_n$ (see exercise 29), it follows that $\sum a_n$ and $\sum b_n$ stand or fall together, as claimed. ∎

How does one pick a "related" series $\sum b_n$ that might yield a useful limit comparison? One way is to think, algebraically, about what the general term of the given series $\sum a_n$ will "look like" in the tail. For instance, consider the series $\sum 1/(2n^3 - n + 1)$. In the tail, the general term $1/(2n^3 - n + 1)$ will look more and more like $1/(2n^3)$, since for huge values of n, the magnitude of $(-n + 1)$ will be dwarfed by that of $2n^3$. With this in mind, we turn to...

Example 2. Test the series $\frac{1}{2} + \frac{1}{15} + \frac{1}{52} + \cdots + \frac{1}{2n^3 - n + 1} + \cdots$ for convergence.

Solution. As discussed above, we'll compare it to $\sum 1/(2n^3)$ with the limit comparison test. Since

$$\lim_{n \to \infty} \left(\frac{1}{2n^3 - n + 1}\right) \Big/ \frac{1}{2n^3} = \lim_{n \to \infty} \frac{2n^3}{2n^3 - n + 1} = 1\,,$$

the two series stand or fall together. Thus, since $\sum 1/(2n^3)$ converges (do you see why?), the given series must converge, too.* ♦

Please bear in mind that the comparison tests apply only to *positive* series, meaning series whose terms are (almost) all positive. [Reread the footnote on this section's first page to explain that "almost"!] Exercise 30 below asks you for counterexamples showing that the comparison tests can fail if you try to apply them to non-positive series.

* The answer to the parenthetical question: $\sum 1/(2n^3) = (1/2)\sum 1/n^3$ (recall exercise 29!). That last series is a p-series, and since p=3, it converges (by Claim 1 above).

Exercises.

30. As stated several times above, the comparison tests work only for *positive* series (i.e. those with a tail of strictly positive terms). In this exercise, you'll consider why this requirement is necessary in the direct comparison tests.

a) To demonstrate that the direct comparison test for convergence does not hold for nonpositive series, think of an example of a nonpositive series $\sum a_n$ that stays below a convergent series $\sum c_n$ and still diverges.

b) To demonstrate that the direct comparison test for divergence does not hold for nonpositive series, think of an example of a nonpositive series $\sum a_n$ that stays above a divergent series $\sum c_n$ and still converges.

c) Demonstrating that the limit comparison test doesn't hold for nonpositive series is harder. We'll return to this in exercise 38, after you've learned Leibniz' alternating series test.

31. State whether each of the following twenty series converges or diverges. Justify your conclusions.

a) $\sum \frac{1}{n^5}$ b) $\sum \frac{1}{\sqrt{n}}$ c) $\sum \left(\frac{3}{4}\right)^{-n}$ d) $\sum \frac{1}{\sqrt[4]{n^3}}$ e) $\sum \frac{1}{\sqrt[3]{n^4}}$ f) $\sum \left(\frac{2}{3}\right)^{n+1}$

g) $\sum \frac{1}{n^2+5}$ h) $\sum \frac{1}{2n^2-5}$ i) $\sum \frac{5}{n}$ j) $\sum \frac{1}{n2^n}$ k) $\sum \frac{n}{n\sqrt{n}-1}$ l) $\sum \frac{n^2}{\sqrt{n^7+1}}$

m) $\sum \frac{1}{n^2+\cos n}$ n) $\sum 3^{1/n}$ o) $\sum \frac{1}{2^{n^2}}$ p) $\sum \frac{1}{n^n}$ q) $\sum \frac{3n^2-12n}{5n^6+3n^2-17}$

r) $\sum \sin(1/n^2)$ s) $\frac{1}{5}+\frac{1}{10}+\frac{1}{15}+\frac{1}{20}+\cdots$ t) $\sum \frac{1}{\ln n}$ [*Hint: How does* $\ln n$ *compare to* n?]

32. Criticize the following argument: In the two series below, each term in the left series is less than its corresponding term in the right series, which is a convergent geometric series. Hence, the left series converges, too.

$$\frac{1}{2}+\frac{1}{5}+\frac{1}{8}+\frac{1}{11}+\cdots \quad \text{compared to} \quad 1+\frac{1}{2}+\frac{1}{4}+\frac{1}{8}+\cdots$$

33. Shortly after you were weaned, you learned that every real number has a decimal representation that either terminates after a finite number of decimal places or rattles on interminably.

a) Every infinite (non-terminating) decimal representation is in fact an infinite series. Explain why.

b) Recall that a number is said to be **rational** if can be represented as a *ratio* of integers. Explain why every terminating decimal represents a rational number.

c) Terminating decimals are always rational, but what about infinite decimals? A clever trick shows that every *repeating* infinite decimal (i.e. one with a string of digits that eventually repeats itself endlessly) is rational. We apply the trick to the decimal $c = 0.818181...$ as follows. Since $100c = 81.818181...$, subtraction yields $99c = 81$. Hence, $c = 9/11$, and is thus a rational number. Make up a repeating decimal, and then use this trick, or some slight variation thereof, to demonstrate that it is rational.

34. The limit comparison test concerns two positive series $\sum a_n$ and $\sum b_n$with the property that (a_n/b_n) approaches a *nonzero* number L in the limit. In this exercise, we'll consider what happens if the ratio approaches zero.

a) Suppose $(a_n/b_n) \to 0$ and $\sum b_n$ converges. In this case, all's well, and $\sum a_n$ converges, too. Explain why. [*Hint: Think about the relative sizes of* a_n *and* b_n *in the limit. Then think about the direct comparison test.*]

b) Suppose $(a_n/b_n) \to 0$ and $\sum b_n$ diverges. In this case, the test fails; we can't conclude anything about $\sum a_n$. To verify that this is so, first consider $a_n = 1/n^2$ and $b_n = 1/n$, and then consider $a_n = 1/n$ and $b_n = 1/\sqrt{n}$.

Alternating Series, Absolute Convergence

"From this hour I ordain myself loos'd of limits and imaginary lines,
Going where I list, my own master total and absolute..."

- Walt Whitman, "Song of the Open Road"

Nonpositive series elude the comparison tests, but occur frequently in practice. The most common type of nonpositive series is an **alternating series**, one in which consecutive terms always have opposite signs. For example, $1 - 1/3! + 1/5! - 1/7! + \cdots$ is an alternating series, which should look familiar to you from your work earlier with Taylor polynomials earlier in this chapter.

In 1705, Leibniz identified a simple convergence test for alternating series.

Leibniz' Alternating Series Test. Let $\sum a_n$ be an alternating series. If $|a_n| \to 0$ *monotonically* (meaning that $|a_{n+1}| < |a_n|$ for all n), then $\sum a_n$ converges.

The figure at right suggests why Leibniz' test works. Because the terms' signs alternate, the series' running total (i.e. its nth partial sum, s_n) hops back and forth on the number line, changing direction with each successive hop. Because the terms' magnitudes vanish monotonically, the hops grow shorter and shorter, and the series must ultimately converge to some number less than all the odd-indexed partial sums (the solid dots in the figure), and greater than all the even-indexed partial sums (the hollow dots).

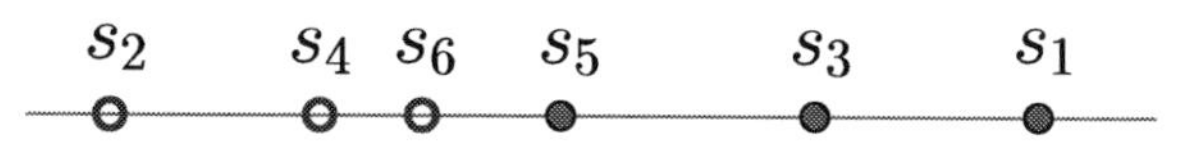

Example 1. The so-called *alternating harmonic series* is $1 - \frac{1}{2} + \frac{1}{3} - \frac{1}{4} + \cdots$. Test it for convergence.

Solution. The general term's magnitude, $1/n$, obviously vanishes monotonically in the series' tail. Hence, by Leibniz' test, the alternating harmonic series converges. ♦

The alternating harmonic series converges, but gracelessly. The convergent series we've met heretofore have all been vigorous and athletic, converging entirely because of the speed with which their terms vanished. Such vigor is sadly absent in the alternating harmonic series. While studying its sibling, the (true) harmonic series, we've already seen that its terms' magnitudes do *not* vanish rapidly enough to converge. Instead, the alternating harmonic series only manages to converge because of the "drag" supplied by its negative terms. It's as if a strong gust of wind, blowing in just the right direction, has helped the alternating harmonic series kick a field goal that it couldn't have managed on its own.

I point this out not to belittle the alternating harmonic series, but to draw an important distinction. We say that a series **converges absolutely** if it converges solely because of the speed with which its terms' magnitudes vanish. (Formally, we say that $\sum a_n$ is absolutely convergent if the series $\sum |a_n|$ converges.) On the other hand, we say that a series **converges conditionally** if it converges only with the help of negative "winds". (Formally, $\sum a_n$ is conditionally convergent if it converges, but $\sum |a_n|$ diverges.)

Absolute convergence is a strong condition. It's worth making explicit (and proving in a footnote) what the name "absolute convergence" correctly implies: If a series converges absolutely, it converges period.* That being so, one way to show that a nonpositive series converges is to show that it converges absolutely.

Example 2. Test the following series for convergence: $1 - \frac{1}{4} - \frac{1}{9} + \frac{1}{16} - \frac{1}{25} - \frac{1}{36} + \frac{1}{49} - \cdots$.

Solution. If its negative terms were made positive, it would be a convergent p-series. Since the original series is absolutely convergent, it is, *a fortiori*, convergent. ♦

Absolutely convergent series are robust. One can prove (as you'll see if you study real analysis) that you can do "natural" things with absolutely convergent series that you can't do with conditionally convergent series. For example, if you reorder the terms in an absolutely convergent series, the sum itself will remain the same, but reordering the terms in a *conditionally* convergent series can change the sum! For example, consider our stock example of conditionally convergence, the alternating harmonic series. If we reorder its terms so that we do two subtractions after each addition, we find that

$$\begin{aligned} &1 - \frac{1}{2} - \frac{1}{4} + \frac{1}{3} - \frac{1}{6} - \frac{1}{8} + \frac{1}{5} - \frac{1}{10} - \frac{1}{12} + \cdots \\ &= \left(1 - \frac{1}{2}\right) - \frac{1}{4} + \left(\frac{1}{3} - \frac{1}{6}\right) - \frac{1}{8} + \left(\frac{1}{5} - \frac{1}{10}\right) - \frac{1}{12} + \cdots \\ &= \frac{1}{2} - \frac{1}{4} + \frac{1}{6} - \frac{1}{8} + \frac{1}{10} - \frac{1}{12} + \cdots \\ &= \frac{1}{2}\left(1 - \frac{1}{2} + \frac{1}{3} - \frac{1}{4} + \frac{1}{5} - \frac{1}{6} + \cdots\right). \end{aligned}$$

Look closely at that chain of equalities: The sum of the rearranged alternating harmonic series is half of what it was before we rearranged it!

Conditional convergence is a fascinating and disturbing phenomenon. Its strangeness is exemplified by the so-called *Riemann Rearrangement Theorem*: Given any conditionally convergent series, we can reorder its terms so that the rearranged series will converge to *any value we like*; we can also reorder its terms in such a way that it will diverge.

A conditionally convergent series is an elaborate house of cards. Breathe on it and it collapses.

In contrast, an absolutely convergent series is safe around small children, and vice-versa. You can push and pull an absolutely convergent series around, rearrange its terms, or even bash your younger brother in the head with it without doing any serious damage. This will be especially important for us when we return to our beloved Taylor series. Taylor series converge, for the most part, absolutely, which means that we'll be able to play fast and loose with them, manipulating them with abandon. We'll even be able to differentiate and integrate them term by term, which will lead us to some interesting results in this chapter's final section.†

---

* *Slick proof*: If $\sum |a_n| = L$, then $\sum(a_n + |a_n|) \leq \sum 2|a_n| = 2L$, so $\sum(a_n + |a_n|)$ converges to some number M. Hence, $\sum a_n = \sum(a_n + |a_n| - |a_n|) = \sum(a_n + |a_n|) - \sum|a_n| = M - L$. That is, $\sum a_n$ converges, as claimed.

† Our ability to differentiate and integrate Taylor series term by term is guaranteed not by absolute convergence, but by the subtler notion of *uniform* convergence, which you can learn about in a real analysis course.

Exercises.

35. For each of the following series, state whether we can conclude that it converges *on the basis of Leibniz' test*. If the answer is no, explain why Leibniz' test does not apply.

a) $1-\frac{1}{3}+\frac{1}{5}-\frac{1}{7}+\cdots$ b) $1-\frac{1}{\sqrt{2}}+\frac{1}{\sqrt{5}}-\frac{1}{\sqrt{10}}+\cdots$ c) $1+\frac{1}{2^2}+\frac{1}{3^2}+\frac{1}{4^2}+\cdots$

d) $\frac{1}{3}-\frac{2}{5}+\frac{3}{7}-\frac{4}{9}+\cdots$ e) $\frac{1}{2^2}-\frac{1}{3^2}+\frac{1}{2^3}-\frac{1}{3^3}+\cdots$ f) $1-\frac{x^2}{2!}+\frac{x^4}{4!}-\frac{x^6}{6!}+\cdots$, for real x.

36. Convince yourself that the series in exercise 35a can be expressed with sigma notation as follows:

$$\sum_{n=0}^{\infty}\frac{(-1)^n}{(2n+1)}.$$

a) Next, express the same series in sigma notation with the index beginning at $n=1$ rather than at 0.

b) Express the series in exercises 35b, d, and f in sigma notation.

37. Find the Maclaurin series for $\ln(1+x)$ and then evaluate it at $x=1$.

38. Classify each of the following series as absolutely convergent, conditionally convergent, or divergent.

a) $\sum(2/3)^n$ b) $\sum(-1/5)^n$ c) $\sum(-1)^n/n^3$ d) $\sum(-1)^n/\sqrt{3n+1}$ e) $\sum(2n(-1)^n)/(5n+1)$

f) $\sum(-1)^{n+1}/\ln(n+3)$ g) $\sum(-1)^{n+1}\sin(n\pi)$ h) $\sum(n^2+42)/20{,}000n^3$ i) $\sum(-1)^n$

39. (Monotonicity in Leibniz' Test) Consider the series

$$\mathbf{1}-\frac{1}{2}+\frac{\mathbf{1}}{\mathbf{2}}-\frac{1}{4}+\frac{\mathbf{1}}{\mathbf{3}}-\frac{1}{8}+\frac{\mathbf{1}}{\mathbf{4}}-\frac{1}{16}+\frac{\mathbf{1}}{\mathbf{5}}-\frac{1}{32}+\cdots,$$

which braids together the divergent harmonic series and a convergent geometric series.

a) Although this is an alternating series whose terms approach zero, Leibniz' test does *not* apply here. Why not?

b) Verify that the $2n^{th}$ partial sum of the series is

$$s_{2n}=\sum_{i=1}^{n}\frac{1}{i}-\sum_{i=1}^{n}\left(\frac{1}{2}\right)^i.$$

c) *If* the series converges to a number s, then the limit of the partial sum above would be s. But then we'd have

$$s=\lim_{n\to\infty}\left(\sum_{i=1}^{n}\frac{1}{i}-\sum_{i=1}^{n}\left(\frac{1}{2}\right)^i\right)=\lim_{n\to\infty}\left(\sum_{i=1}^{n}\frac{1}{i}\right)-\lim_{n\to\infty}\left(\sum_{i=1}^{n}\left(\frac{1}{2}\right)^i\right)$$

$$=\left(\sum_{i=1}^{\infty}\frac{1}{i}\right)-\left(\sum_{i=1}^{\infty}\left(\frac{1}{2}\right)^i\right)=\left(\sum_{i=1}^{\infty}\frac{1}{i}\right)-1.$$

Adding 1 to both sides would then yield the absurd conclusion that the harmonic series *converges* (to $s+1$). That being the case, what can we conclude?

40. (Unfinished business from exercise 30c) To demonstrate that the limit comparison test *doesn't* hold if the series involved are *not* positive series, consider the following series:

$$\sum a_n=\sum\left(\frac{(-1)^n}{\sqrt{n}}+\frac{1}{n}\right) \quad \text{and} \quad \sum b_n=\sum\frac{(-1)^n}{\sqrt{n}}$$

a) Verify that $(a_n/b_n)\to 1$ and that $\sum b_n$ converges.

b) Despite the previous part, explain why $\sum a_n$ diverges.

c) Remember the moral: The comparison tests (including the limit comparison test) apply only to *positive* series.

The Ratio Test

"In the beginning was the *ratio*..."
- John 1:1

With a satisfying sense of poetry, our last major convergence test returns us to our first.

In a geometric series, the ratio of any term to its predecessor is some fixed value, r. If the magnitude of this term-to-predecessor ratio is less than 1, the geometric series converges. Otherwise, it diverges.

In a non-geometric series, the term-to-predecessor ratio isn't fixed, but it may still ***approach*** a fixed value in the tail of the series, where convergence is determined. If this happens, then the non-geometric series eventually becomes indistinguishable from a geometric series. From this simple insight is born the so-called ratio test.

Ratio Test. Let $\sum a_n$ be a series. Consider the magnitude of its term-to-predecessor ratio, $\left|\frac{a_n}{a_{n-1}}\right|$.
If, in the limit, this approaches a number less than 1, then the series converges absolutely.
If, in the limit, this approaches a number greater than 1, then the series diverges.
If, in the limit, this approaches 1 itself, then the ratio test is inconclusive.

In the exercises, you'll justify the statement that the ratio test is inconclusive if the term-to-predecessor ratio approaches 1 in absolute value. For now, here are a couple of examples of the ratio test in action.*

Example 1. Test the series $\sum n^3/3^n = \frac{1}{3} + \frac{8}{9} + 1 + \frac{64}{81} + \cdots$ for convergence.
Solution. The ratio of the general term to its predecessor is, in absolute value,

$$\left|\frac{\frac{n^3}{3^n}}{\frac{(n-1)^3}{3^{n-1}}}\right| = \frac{1}{3}\left|\frac{n}{n-1}\right|^3, \text{ which approaches } 1/3 \text{ as } n \to \infty.$$

Hence, by the ratio test, the series is absolutely convergent. ♦

Example 2. Test the series $\sum 2^n/\sqrt{n}$ for convergence.
Solution. The ratio of the general term to its predecessor is, in absolute value,

$$\left|\frac{\frac{2^n}{\sqrt{n}}}{\frac{2^{n-1}}{\sqrt{n-1}}}\right| = 2\sqrt{\frac{n-1}{n}}, \text{ which approaches } 2 \text{ as } n \to \infty.$$

Hence, by the ratio test, the series diverges (which is also obvious from the divergence test). ♦

The ratio test is especially useful when we are working with infinite-degree polynomials, to which we'll return at long last in the next section. But first, exercises.

---

* Since the ratio test involves absolute values, it's worth remembering that $|ab| = |a||b|$ and that $|a/b| = |a|/|b|$. These two simple properties of the absolute value often come in handy when we use the ratio test.

Exercises.

41. Each of the following represents the general term of a series $\sum a_n$. Write down an algebraic expression for the general term's predecessor, a_{n-1}.

a) $a_n = n^2$ b) $a_n = 2n$ c) $a_n = n(n+1)$ d) $a_n = 2n(2n+1)$ e) $a_n = 5^{3n-1}/n!$

42. When using the ratio test, it's not uncommon to end up juggling factorials in the term-to-predecessor ratio. To practice your juggling skills, simplify the following expressions.
[*Hint: When in doubt, write it out.* (*That is, write $n!$ out, or at least* ***think*** *it out, as* $2 \cdot 3 \cdot 4 \cdots n$)]

a) $8!/6!$ b) $n!/(n+1)!$ c) $\dfrac{(2n)!}{(2n-3)!}$ d) $\dfrac{n!}{(2n+1)!} \cdot \dfrac{(2n-1)!}{(n-1)!}$

43. Test each of the following series for convergence by using the ratio test.

a) $\sum \dfrac{n}{10^{n-1}}$ b) $\dfrac{3}{5} + \dfrac{4}{5^2} + \dfrac{5}{5^3} + \cdots$ c) $\sum \dfrac{2^n}{n}$ d) $\dfrac{1}{10} + \dfrac{2!}{100} + \dfrac{3!}{1000} + \dfrac{4!}{10{,}000} + \cdots$

e) $\dfrac{1}{3} + \dfrac{2!}{3 \cdot 5} + \dfrac{3!}{3 \cdot 5 \cdot 7} + \dfrac{4!}{3 \cdot 5 \cdot 7 \cdot 9} + \cdots + \dfrac{n!}{3 \cdot 5 \cdot 7 \cdot 9 \cdots (2n+1)} + \cdots$ f) $\sum \dfrac{3}{(4n-1)4n}$

44. Above, I claimed that the ratio test tells us nothing if the term-to-predecessor ratio goes to 1. To see that I'm not lying to you, show that in the general p-series, $\sum 1/n^p$, this ratio always approaches 1 in the limit, regardless of the value of p. Then explain why this fact justifies my statement.

45. In many calculus books, the ratio given in the ratio test is $\left|\dfrac{a_{n+1}}{a_n}\right|$ rather than $\left|\dfrac{a_n}{a_{n-1}}\right|$.
Explain why this other form works just as well. Then try it out by redoing a few parts of exercise 43.

46. The so-called **root test** is similar in spirit to the ratio test. To use it on a series $\sum a_n$, take the limit of $\sqrt[n]{|a_n|}$ as $n \to \infty$. If the limit is less than 1, the series converges absolutely; if the limit exceeds 1, the series diverges.

a) Explain intuitively why the root test works. [*Hint: If* $\sqrt[n]{|a_n|} \to r$, *then* $|a_n| \to r^n$. *Now think about the tail.*]

b) Use the root test to test the following series for convergence: i) $\sum \left[\dfrac{5n^2-1}{4n^2+10n}\right]^n$ ii) $\sum \dfrac{1}{(\ln n)^n}$

47. A few sections back, we established the convergence of p-series (when $p > 1$) by comparing them to certain improper *integrals*. If we generalize the technique we used, we are led to the following **integral test**:

> If $\sum a_n$ is a series of positive terms such that $a_n = f(n)$ for some decreasing function f, then *The series $\sum a_n$ and the improper integral $\int_1^\infty f(x)\,dx$ either both converge or both diverge.*

a) With the help of the two figures below, explain *why* the integral test holds.

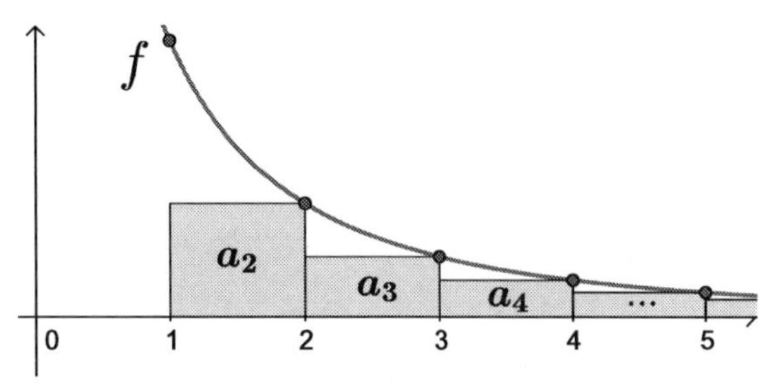

b) Apply the integral test to the following series: i) $\sum n e^{-n^2}$ ii) $\sum (\ln n)/n^2$ [*Hint: Integration by parts.*]

Power Series

Dachshunds have specific (and quite funny) anatomical peculiarities, but the best way to understand most features of dachshund anatomy is to learn canine anatomy in general. Similarly, the best way to learn about most features of Taylor series – one "breed" of infinite-degree polynomial – is to study the features common to *all* infinite-degree polynomials. The term "infinite-degree polynomial", incidentally, is considered a bit vulgar by the mathematician in the street, whose preferred term for an infinite-degree polynomial is the shorter, but less expressive "power series".

Formally, a **power series** in x (centered at zero) is any function of the form

$$a_0 + a_1 x + a_2 x^2 + a_3 x^3 + \cdots \ .^*$$

Substituting a value for x turns a power series into a series of constants – which may converge or diverge. Using the ratio test, we can determine most of the values of x which make a given power series converge; a little fine tuning with other convergence tests will then finish the job.

Example 1. For which values of x does the power series $\sum \frac{n}{4^n} x^n$ converge?

Solution. For any given x, the term-to-predecessor ratio of this series is, in absolute value,

$$\left| \frac{nx^n}{4^n} \cdot \frac{4^{n-1}}{(n-1)x^{n-1}} \right| = \frac{n}{4(n-1)} |x|.$$

As n goes to infinity, this ratio goes to $|x|/4$. This is less than 1 when $|x| < 4$, so by the ratio test, the power series *converges absolutely* whenever $-4 < x < 4$. The ratio test also tells us that the power series diverges when $|x| > 4$, but it is inconclusive when $|x| = 4$. Thus, so far, we know that the series converges throughout the open interval $(-4, 4)$, and it *might* converge at the endpoints of the interval, $x = 4$ and $x = -4$. We'll need to check these endpoints separately.

When $x = 4$, the series $\sum (nx^n)/4^n$ becomes $\sum n$, which diverges. When $x = -4$, it becomes $\sum (-1)^n n$, which diverges, too. Hence, the power series converges only when $\boldsymbol{-4 < x < 4}$. ♦

We'll soon prove that *every* power series converges for all x within some **interval of convergence** specific to that series, and diverges for all x outside of it. From this interval's midpoint (which is the value at which the series itself is centered), the interval extends the same distance to the right and left. This distance is called the **radius of convergence** for the power series. For instance, the example above concerns a power series centered at zero; its interval of convergence is $(-4, 4)$, and its radius of convergence is 4.

Example 2. Find the interval of convergence for $\sum x^n/n!$, the Maclaurin series for e^x.

Solution. For any value of x, this series' term-to-predecessor ratio is, in absolute value,

$$\left| \frac{x^n}{n!} \cdot \frac{(n-1)!}{x^{n-1}} \right| = \frac{|x|}{n},$$

which goes to zero as $n \to \infty$. Hence, by the ratio test, this series converges absolutely for *all* x. The interval of convergence is thus $(-\infty, \infty)$; the radius of convergence is ∞. ♦

---

* More generally, a power series **centered at $\boldsymbol{a}$** has the form $a_0 + a_1(x-a) + a_2(x-a)^2 + a_3(x-a)^3 + \cdots$.

Example 3. Find the interval of convergence of the power series $\sum \frac{3^n}{n}(x-2)^n$.

Solution. For any value of x, this series' term-to-predecessor ratio is, in absolute value,

$$\left|\frac{3^n(x-2)^n}{n} \cdot \frac{n-1}{3^{n-1}(x-2)^{n-1}}\right| = \frac{3(n-1)}{n}|x-2|,$$

the limit of which is $3|x-2|$. This is less than 1 when $|x-2| < 1/3$, so by the ratio test, the series converges absolutely when x is within $1/3$ unit of 2. (That is, when $5/3 < x < 7/3$.)

As always, we must investigate the endpoints separately.

As you should verify, when $x = 7/3$, the series becomes the harmonic series, so it diverges. When $x = 5/3$, it becomes the *alternating* harmonic series, so it converges.

Hence, the interval of convergence is $\mathbf{[5/3\,, 7/3)}$. ♦

In each of the three preceding examples, we've seen a power series converge on an interval, but three confirmations do not constitute proof – except in the works of Lewis Carroll.* One can confirm, after all, that 3, 5, and 7 are all prime numbers, but this hardly proves that all odd numbers are prime.

Theorem. Let $\sum a_n(x-a)^n$ be a power series centered at a. The values of x that make it converge constitute an interval centered at a. At all points in its interior, the series converges absolutely.

Proof. The idea is to show that if the series converges at some number c, then it must also converge absolutely for all numbers *closer* to a than c is. Read no further until you are convinced that by proving this, we will indeed have proved the theorem.

Suppose the series converges when $x = c$. Suppose b is any number closer to a than c is.

Because b is closer to a than c is, the geometric series $\sum|(b-a)/(c-a)|^n$ converges.

We'll now use a direct comparison test to show that if we evaluate the power series at b, the resulting numerical series converges absolutely. To begin, we'll decompose the general term of this numerical series (in absolute value) as follows:

$$|a_n(b-a)^n| = |a_n(c-a)^n|\left|\frac{(b-a)^n}{(c-a)^n}\right|.$$

Since $\sum a_n(c-a)^n$ converges, the decomposition's first factor, $|a_n(c-a)^n|$, vanishes in the limit. Consequently, in the tail of $\sum|a_n(b-a)^n|$, the general term will be the product of two things: (1) something vanishingly small, and (2) the decomposition's second factor. Hence, this general term will always be significantly *less than* the second factor. But the second factor is the general term of the convergent geometric series we identified earlier. Hence, by direct comparison, the series $\sum|a_n(b-a)^n|$ converges. Thus, $\sum a_n(x-a)^n$ converges *absolutely* at b, as claimed. ■

---

* *"Just the place for a Snark!" the Bellman cried,*
As he landed his crew with care;
Supporting each man on top of the tide
With a finger entwined in his hair.

"Just the place for a Snark! I have said it twice:
That alone should encourage the crew.
Just the place for a Snark! I have said it thrice:
What I tell you three times is true."

- Lewis Carroll, "The Hunting of the Snark".

Exercises.

48. To work with power series, you must be comfortable with combinations of inequalities and absolute values, such as $|x - a| < b$. You should be able to think about such things geometrically, and work with them algebraically. Some practice: if x, a and b are real numbers...

a) Define $|x|$ algebraically.
b) What does $|x|$ represent geometrically?
c) Explain why $|a - b|$ represents the distance between a and b on the number line.
d) Describe, geometrically, the set of all real numbers x satisfying the inequality $|x| < a$.
e) Describe, geometrically, the set of all real numbers x satisfying the inequality $|x - a| < b$.
f) Explain why $|x| < a$ is equivalent to $-a < x < a$.
g) Explain why $|x - a| < b$ is equivalent to $-b < x - a < b$, and thus to $a - b < x < a + b$.
h) Rewrite the following *without* absolute value signs: $|x| < 2$, $|x - 3| \leq 2/3$, $|2x - 1| < 5$.

49. Can a power series' interval of convergence be $(0, \infty)$? Why or why not?

50. For each of the following power series, find the interval of convergence.

a) $\sum \frac{(-1)^{n+1}}{n^2} x^n$ b) $\sum \frac{1}{n^3 + 1} x^n$ c) $\sum \frac{1}{n + 4} x^n$ d) $\sum \frac{1}{(-5)^n} x^{2n+1}$ e) $\sum \frac{(-3)^n}{n} x^n$

f) $\sum \frac{n^2}{2^n} (x - 4)^n$ g) $\sum \frac{x^{2n}}{(2n)!}$ h) $\sum (-1)^{n+1} n x^{n-1}$ i) $\sum \frac{(x-2)^n}{n!}$ j) $\sum \frac{\ln n}{e^n} (x - e)^n$

51. It is possible for a power series to have a radius of convergence of *zero*. Such a series is worthless in practice, since it is a function whose domain consists of a single point.

a) Verify that the power series $\sum (n!) x^n$ converges *only* when $x = 0$.
b) Is it possible for a power series to diverge at every real number?

52. a) I stated earlier that a Taylor series is a special "breed" of power series. Explain why this is true.

b) Find the intervals of convergence for the Maclaurin series of $\sin x$, $\cos x$, $1/(1 - x)$, and $\ln(1 + x)$.

53. (Close, but no cigar.) The proof of the theorem on the preceding page is sneaky; surely there is a cleaner proof? Alas, if there is – which I doubt – I don't know it. Here's a near-miss, however: a natural approach that *almost* works, but doesn't quite. Your job is to spot its fatal flaw.

> **Theorem.** Let $\sum a_n (x - a)^n$ be a power series centered at a. The values of x that make it converge constitute an *interval* centered at a. At all points in the interval's interior, the series converges *absolutely.*
>
> **"Proof".** We'll show that the series converges at c, then it converges absolutely for all numbers *closer* to a than c is. So, suppose $\sum a_n (c - a)^n$ converges, and let b be such that $|b - a| < |c - a|$. From this inequality, it follows that $|a_n (b - a)^n| < |a_n (c - a)^n|$ for all n. Hence, by direct comparison with $\sum |a_n (c - a)^n|$, which is known to be convergent, we conclude that $\sum |a_n (b - a)^n|$ converges.
> Hence, $\sum a_n (b - a)^n$ converges *absolutely*, as claimed. ☠

Spot the flaw in this supposed proof.

(Five Terms in) A Series of Stories of Series

Lest the magic of series be obscured by technical details, I'll end this chapter with a few real (and complex) showstoppers, leaving careful justifications to another course. Scandalized professors should instruct their students to shut their eyes and plug their ears. For the rest of you, enjoy.

Story 1. With patience, any poor drudge who knows how to take derivatives can construct, term by term, a Taylor series for any elementary function. Those with vulpine eyes, however, can sometimes spot a craftier method: Hack an already-familiar series. For example, start with your old friend

$$\frac{1}{1-x} = 1 + x + x^2 + x^3 \cdots \qquad \text{(when } |x| < 1\text{)}.$$

Now substitute $-8x^3$ for x, and you'll obtain a Taylor expansion of an entirely different function:

$$\frac{1}{1+8x^3} = 1 - 8x^3 + 64x^6 - 512x^9 + \cdots \qquad \text{(when } |-8x^3| < 1\text{)}.^*$$

Thus, without having to compute derivatives, we've found a Taylor series for $1/(1+8x^3)$. We even know its radius of convergence.

Similarly, substituting $-x^2$ for x in the series for $1/(1-x)$ immediately yields

$$\frac{1}{1+x^2} = 1 - x^2 + x^4 - x^6 + \cdots \qquad \text{(when } |x| < 1\text{)}.$$

Ah, but watch what we can do next! Behold – the series expansion of *arctangent*:

$$\arctan x = \int \frac{1}{1+x^2}dx = \int (1 - x^2 + x^4 - x^6 + \cdots)dx = x - \frac{x^3}{3} + \frac{x^5}{5} - \frac{x^7}{7} + \cdots.$$

It gets even better. If we let $x = 1$ in the leftmost and rightmost expressions, we find that

$$\boldsymbol{\frac{\pi}{4} = 1 - \frac{1}{3} + \frac{1}{5} - \frac{1}{7} + \cdots}.$$

This series, which links π to the sequence of odd numbers, converges far too slowly to help us approximate π's numerical value.[†] But who cares? The series itself is beautiful – far more beautiful than π's random-looking decimal expansion.[‡] It has long been called "Leibniz's series" because Leibniz discovered it in the early 1670's, though unbeknownst to him, it had appeared in print a few years earlier in the work of Scottish mathematician James Gregory. As it turns out, "Leibniz' series" had also been discovered a few *centuries* earlier in the remarkable "Kerala school of mathematics", the existence of which was unknown outside of southern India until the mid-19th century.

[*] The condition $|-8x^3| < 1$ reduces to $|x| < 1/2$ as you should verify.

[†] Even after summing 100 terms of the series (and multiplying by 4), the result is still closer to 3.13 than 3.14. To obtain even 10 decimal places of accuracy, one would need to compute over 5 *billion* terms of the series.

[‡] Moreover, a real number's decimal expansion has no intrinsic mathematical significance. Our use of base ten is just a biological accident. Had evolution given our species just four fingers on each hand, we'd likely be using base eight. Long ago, you would have learned the beginning of π's *octal* expansion ($\pi = 3.1103755$...), and if a strange textbook author were to include π's *decimal* expansion in a footnote, you would find it uncomfortably alien.

Story 2. In 1734, Euler linked the circular constant π to the sequence of squares by proving that

$$1+\frac{1}{4}+\frac{1}{9}+\frac{1}{16}+\frac{1}{25}+\cdots=\frac{\pi^2}{6}.$$

Euler's proof settled a famous problem that had thwarted generations of mathematicians in the 90 years since it had first been posed: Find the exact sum of the p-series when $p = 2$.

Euler being Euler, in the years that followed, he found the exact sums of many other p-series. To call Euler a voluminous writer is like calling a Giant Sequoia a large plant. The statement is true, but underplays a magnitude that is well-nigh unbelievable. The definitive publication of his *Opera Omnia* (complete works) began in 1911, with a new volume appearing every few years. Over a century later, the collection is still not quite complete.* Should you ever have the opportunity to stand before the shelves containing Euler's (almost) complete works at a good academic library, do so. It is a humbling experience. To read Euler's actual words therein, you'll need to know at least one of the languages in which he wrote: Latin, French, German, and Russian.

In 1750, Euler established a formula that gives the exact sum of the p-series for any *even* value of p whatsoever.† Curiously, the odd values of p proved entirely resistant even to Euler's genius. Never in his long life could he find the exact sum of any p-series for which p is odd. And today, almost a quarter of a millennium after Euler's death, no substantial progress has been made on that problem. What is the exact sum of $\sum 1/n^3$? No one knows.‡

A deep twist in the p-series story appeared in the 19th century work of Bernhard Riemann, whose mathematical style contrasts nicely with Euler's. Riemann's works fit into a single volume, nearly every page of which revolutionized some subfield of mathematics. (His work in differential geometry also paved the way for Einstein's theory of general relativity.) In a paper on number theory, Riemann initiated the study of what is now called the *Riemann zeta function*. I cannot hope to explain here either what the Riemann zeta function is or why it is important, except to say that it is intimately connected to the distribution of prime numbers, and that at its core lies this function:

$$\zeta(s)=\sum_{n=1}^{\infty}\frac{1}{n^s}.$$

As you can see, when s is a real number greater than one, $\zeta(s)$ is the sum of a convergent p-series. The zeta function, however, is defined for all complex values of s, except $s = 1$. The solutions to the equation $\zeta(s) = 0$ are the subject of the so-called "Riemann Hypothesis", the single most famous unsolved problem in pure mathematics today – and in the foreseeable future. Among the many semi-popular books on the Riemann Hypothesis that have been written, I recommend John Derbyshire's *Prime Obsession*. (Also, if you manage to prove the Riemann Hypothesis, a million-dollar prize awaits you, courtesy of the Clay Institute.)

---

* The last volume (comprising papers on perturbation theory in astronomy) is scheduled for publication in 2020. For an accessible introduction to Euler's life and mathematics, see William Dunham's book, *Euler: The Master of Us All*.

† The formula involves the so-called "Bernoulli numbers", which I mentioned in a footnote earlier in this chapter.

‡ Its numerical value is approximately 1.2020569. It is sometimes called "Apery's constant" after Roger Apery, who proved in 1978 that it is an irrational number.

Story 3. To extend the domain of the exponential function to complex values, mathematicians simply let x be complex in the Taylor series expansion of e^x.* This reduces the problem of what is meant by, say e^{2+3i} to a matter of addition and multiplication, which are straightforward operations, even with complex numbers. For example, raising e to the power of a pure imaginary number $i\theta$ (where θ could be any real number) yields

$$e^{i\theta} = 1 + i\theta + \frac{(i\theta)^2}{2!} + \frac{(i\theta)^3}{3!} + \frac{(i\theta)^4}{4!} + \frac{(i\theta)^5}{5!} + \frac{(i\theta)^6}{6!} + \frac{(i\theta)^7}{7!} + \frac{(i\theta)^8}{8!} + \cdots.$$

Each term above involves powers of i. When we raise i to successive powers, a pattern emerges. Play around with this on your own; you'll find that every whole number power of i is one of four numbers: $i, -1, -i$, or 1. After a bit more fooling around, you'll also be able to convince yourself that the series above can be rewritten as

$$e^{i\theta} = 1 + i\theta - \frac{\theta^2}{2!} - i\frac{\theta^3}{3!} + \frac{\theta^4}{4!} + i\frac{\theta^5}{5!} - \frac{\theta^6}{6!} - i\frac{\theta^7}{7!} + \frac{\theta^8}{8!} + \cdots.$$

Written this way, we see that the even-powered terms are real, while the odd-powered terms are imaginary. Separating the real and imaginary terms and factoring the i from the latter reveals that

$$e^{i\theta} = \left(1 - \frac{\theta^2}{2!} + \frac{\theta^4}{4!} - \frac{\theta^6}{6!} + \frac{\theta^8}{8!} + \cdots\right) + i\left(\theta - \frac{\theta^3}{3!} + \frac{\theta^5}{5!} - \frac{\theta^7}{7!} + \cdots\right).$$

Ah! The Taylor series for cosine and sine! We've just established that for any real value of θ,

$$\boldsymbol{e^{i\theta} = \cos\theta + i\sin\theta.}$$

This shocking formula, which relates the exponential function to the trigonometric functions (!), is a central result in the subject of complex analysis, and thus to all branches of mathematics and physics that make use of complex analysis. It is known as *Euler's formula*. If, in Euler's formula, we let θ be π, we obtain $e^{\pi i} = -1$. Adding 1 to both sides of this equation yields one of the most elegant of all equations:

$$e^{\pi i} + 1 = 0,$$

which relates the five most important numbers in as compact a space as possible.

* See chapter 3, exercise 9 for the story of where complex numbers come from in the first place.

Story 4. If you take a course in complex analysis, you'll learn a remarkable geometric connection between a given Taylor series' center and its radius of convergence: Briefly, if f is a function, then the radius of convergence for its Taylor expansion centered at a is the distance from a to the nearest point where f "crashes".

For example, the function $1/(1-x)$ crashes only when $x = 1$, where it is undefined, so the radius of convergence of its Maclaurin series (which is centered at 0 by definition) must be 1, which you already know by other means.

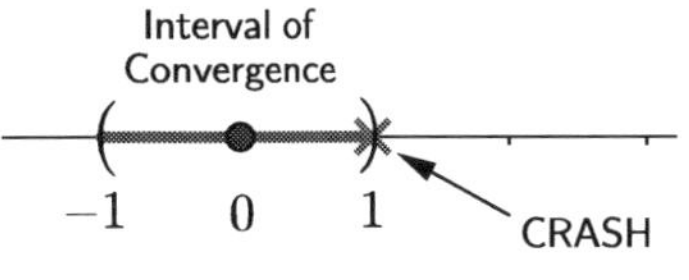

On the other hand, if we were to center the Taylor series for the same function at $x = 3$, the radius of convergence would become 2, since this is the distance from the series' new center to the point where the function crashes.

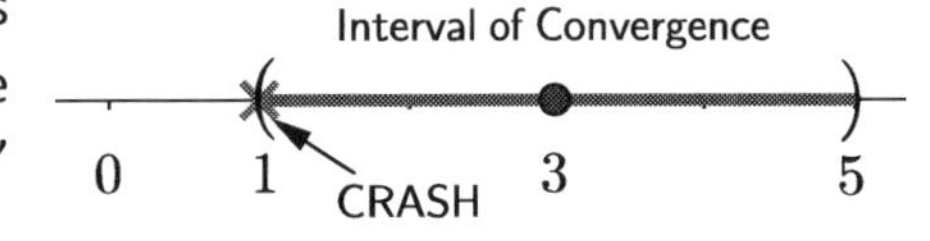

Consider another example: Because e^x never crashes, its Maclaurin series has an *infinite* radius of convergence. The same holds true for the familiar Maclaurin series for $\sin x$ and $\cos x$.

What about the Maclaurin series for $1/(1+x^2)$? Early in this chapter, we observed empirically that this function's Taylor polynomials (centered at 0) *seemed* to approximate it well only when $|x| < 1$. Older and wiser, we could now use the ratio test to prove this conjecture, which is nice, but what we really want is an intuitive explanation as to *why* $1/(1+x^2)$ has such a limited interval of convergence. After all, the function never crashes (it is defined for all reals), so why doesn't its Maclaurin series converge for all reals? The answer is beautiful: The function *does* crash… at i. This "imaginary" crash 1 unit from the origin makes the radius of convergence 1.

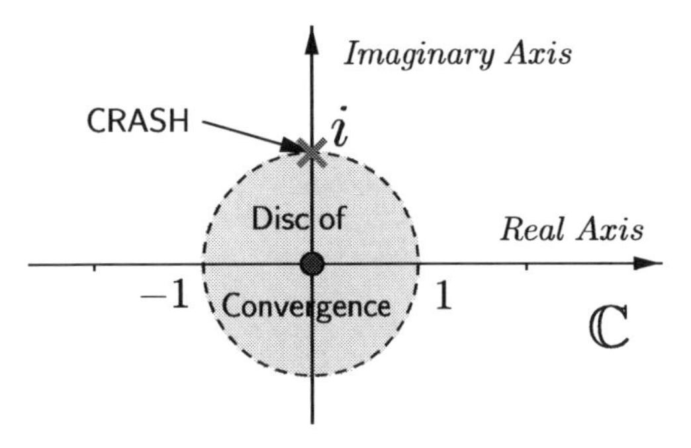

This way of thinking explains the term *radius* of convergence. When we view a power series as being defined throughout the complex plane (rather than just on the real line), it converges for all complex numbers within a particular *disc* – and diverges for all complex numbers outside of it. If the "disc of convergence" happens to intersect the real axis, then from the myopic perspective of one limited to the real numbers, the intersection (which will be an interval, of course) will be the power series' interval of convergence on the real line.

To understand our provincial "real world" deeply, we must sometimes stand outside of it and view it in the wider context of the complex plane.

Story 5. Suppose that after some bizarre catastrophe induced by a mad attempt to divide by zero, mankind has lost nearly all knowledge of trigonometry. One cryptic fragment survives – a reference to a function called *sine*, plus the values of all of its derivatives (including its zeroth derivative) at 0.

From this scanty information, can we deduce anything else about how sine behaves? The prospects might seem bleak: We are told how sine behaves at (and infinitesimally close to) zero, but we lack direct information about how it behaves anywhere else.

And yet… thanks to the miracle of Taylor series, we can reconstruct the entire sine function. This follows immediately from our recipe for a function f's Maclaurin series, $\sum\left(f^{(n)}(0)/n!\right)x^n$, which is determined entirely by f's derivatives at zero!

Exercises.

54. Hack series you know already to find Maclaurin series (and their radii of convergence) for the following functions.

a) $\frac{1}{1+x}$ b) $\frac{1}{1+16x^4}$ c) $\sin(x^2)$ d) e^{-x^2}

55. Find the Maclaurin series for $\ln(1+x)$ by integrating the terms of another series. [*Hint: Use #54a.*]

56. Use the series you found in the previous exercise to find the sum of the alternating harmonic series.

57. Since e^{-x^2} lacks an elementary antiderivative, we can't evaluate $\int_0^1 e^{-x^2}dx$ with the FTC. However, we can easily *approximate* its value; we need only replace the integrand with an appropriate Taylor polynomial and integrate that. Approximate this integral's value by substituting a 6th-degree Taylor polynomial for e^{-x^2} (see exercise 54d).

58. When we widen the playing field from the reals to the complex numbers, we can take square roots of negative numbers. We can also take *logarithms* of negative numbers. As a complex number, what is $\ln(-1)$?

59. Square roots and logarithms of *imaginary* numbers are also possible!

a) Find $\sqrt{i}$. [*Hint: If* $\sqrt{i} = a + bi$, *then* $i = \cdots$.] b) Find $\ln(i)$. [*Hint: Euler's formula.*]

60. Differentiating term by term, verify these statements. Then explain why they make intuitive sense.

a) $\frac{d}{dx}\left(1 + x + \frac{x^2}{2!} + \frac{x^3}{3!} + \cdots\right) = 1 + x + \frac{x^2}{2!} + \frac{x^3}{3!} + \cdots$ b) $\frac{d}{dx}\left(x - \frac{x^3}{3!} + \frac{x^5}{5!} + \cdots\right) = 1 - \frac{x^2}{2!} + \frac{x^4}{4!} + \cdots$

61. Find the radius of convergence for the Maclaurin series of each of the following: a) $\frac{1}{x^2+x-6}$ b) $\frac{3x+2}{x^2+2x+5}$.

62. For each function in problem 61, find the radius of convergence for the Taylor series centered at $x = -2$.

63. Long ago, back in exercise 13f, you found the *binomial series*: For any number α,

$$(1+x)^\alpha = 1 + \alpha x + \frac{\alpha(\alpha-1)}{2!}x^2 + \frac{\alpha(\alpha-1)(\alpha-2)}{3!}x^3 + \cdots.$$

a) If α is a positive integer, the binomial series is not an infinite series, but a finite polynomial. Explain why the right-hand side of the equation will have finitely many terms in this case.

b) If $\alpha = -1$, verify that the binomial series is the same series you found in exercise 54a for $1/(1+x)$.

c) The Maclaurin series for $\arcsin x$ has a peculiar pattern in its coefficients. You can find the series as follows: First, expand the binomial series in the case when $\alpha = -1/2$. Then substitute $-x^2$ for x to obtain a Maclaurin series for the derivative of $\arcsin x$. Integrate this term by term to obtain the series for $\arcsin x$.

64. A *differential equation* (DE) is one that gives a relationship between an unknown function y and its derivative(s). The game is to find a function y that satisfies the DE. Such a function is called a *solution* to the DE.

The simplest DE is $y' = y$. Obviously, any function of the form $y = ce^x$ satisfies this DE (for any real c), but **are there any other solutions**? That is, besides the exponential function (and constant multiples of it), is there another function that equals its own derivative? Power series give us one way of answering our question. The idea is that since any reasonable function is representable by a power series, let's assume that any solution to our DE can be represented as $y = \sum_0^\infty a_n x^n$ for some choice of coefficients $a_0, a_1, a_2, \ldots$. Your problem: Substitute this expression into the DE, and then, by equating coefficients in the resulting equation, prove that a power series solution of $y = y'$ must have the form

$$a_0 + a_0 x + \frac{a_0}{2!}x^2 + \frac{a_0}{3!}x^3 + \cdots.$$

Then explain why this answers the boldface question above in the negative, at least among functions capable of being represented by a power series centered at zero.*

* We can solve $y' = y$ by much simpler means (as a *separable* DE), but the series approach has an odd charm.

Chapter 8

Polar Coordinates and Parametric Equations

Polar Coordinates

"Here, Saturn's grey chaos rolls over me, and I obtain dim, shuddering glimpses into those Polar eternities..."
- Hermann Melville, *Moby Dick*, Ch. 104

When we give each point in the plane a pair of coordinates (an address, essentially), we link the geometric world of points and curves to the algebraic world of numbers and equations. We most frequently use Cartesian (x and y) coordinates, but in some situations, particularly those involving motion or symmetry around a distinguished point, another coordinate system is more appropriate.

In this other coordinate system, we choose an origin, called the *pole*, and a ray emanating from it, called the *polar axis*. Relative to these, each point in the plane can then be given a pair of **polar coordinates**, (r, θ), as indicated at right: r is the point's distance from the pole; θ is measured counterclockwise from the polar axis. When plotting a point (r, θ), it's helpful to think dynamically: Imagine a ray, initially lying along the polar axis, rotating through angle θ; then imagine a point, emitted from the pole, sliding r units along it.

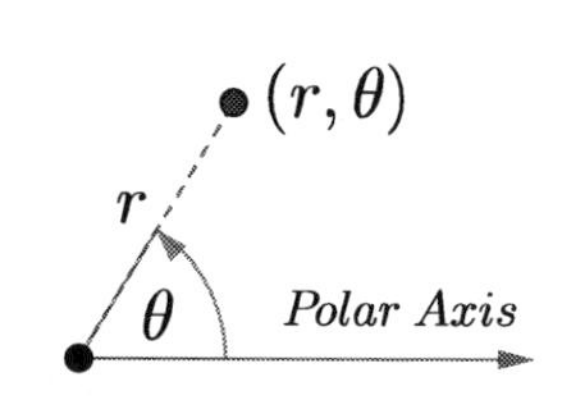

So far so simple, until we notice that each point in the plane has multiple representations in polar coordinates. For example, point P in the figure at right is clearly $(2, \pi/2)$, but it is also $(2, 5\pi/2)$ and $(2, -3\pi/2)$. In fact, each point has infinitely many representations in polar coordinates. This is quite unlike the Cartesian coordinate system, in which each point's representation is unique.

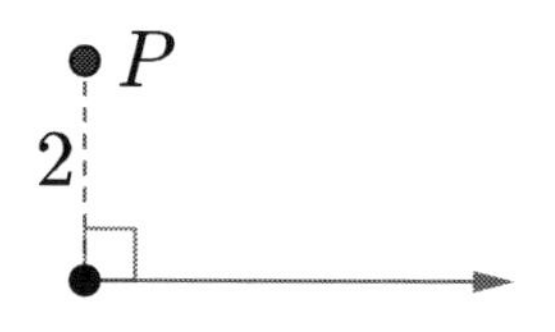

We could eliminate these multiple representations by insisting that $0 \leq \theta < 2\pi$, but this turns out to be more awkward than helpful.* So, we typically leave θ and r unrestricted. Negative values of θ have an obvious meaning, while negative values of r are easy to understand if, as suggested above, we plot our points dynamically. For example, to plot a point such as $(-2, 39°)$, we rotate our ray (initially lying on the polar axis) through $39°$, then we slide our point (initially lying on the pole) -2 units along the ray, meaning 2 units in the direction *opposite* that of the ray, as the figure at right makes clear.

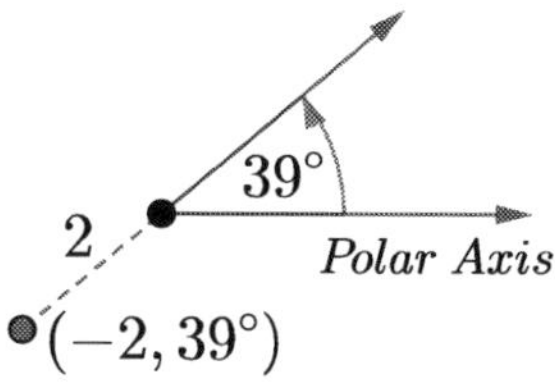

Exercises.

1. Plot the points in the plane given by the following polar coordinate pairs.

$A: (1, \pi/6)$ $B: (2, \pi/3)$ $C: (2, -\pi/2)$ $D: (2, \pi)$ $E: (1, 11\pi/4)$
$F: (2, 300\pi)$ $G: (1, -1)$ $H: (-2, \pi/4)$ $I: (-1, -\pi/2)$ $J: (0, 212°)$

2. The Cartesian coordinates of seven points are given below. Give the polar coordinates for each of these points, subject to the restrictions $r \geq 0$ and $0 \leq \theta < 2\pi$.

a) $(0,3)$ b) $(3,0)$ c) $(-3,0)$ d) $(0,-3)$ e) $(1,1)$ f) $(-2,-2)$ g) $(1, \sqrt{3})$

---

* Compare fractions. Every fraction has multiple representations ($1/2 = 2/4 = 3/6$, etc). Why not, for simplicity's sake, accept only *reduced* fractions (those without common factors on top and bottom)? You'll find out if you try. [For example, explain why $2/3 + 1/6 = 5/6$ without using any non-reduced fractions in your explanation. It can be done, but awkwardly.]

Polar Equations and their Graphs

What is the unit circle's equation? In Cartesian coordinates, everyone knows the answer: $x^2 + y^2 = 1$. However, the equation becomes much simpler if we drop Descartes and think instead like polar bears. The unit circle consists of all points whose r-coordinate is 1. Hence, its polar equation is simply $\boldsymbol{r = 1}$. As this example shows, polar coordinates dramatically simplify some curves' equations.

Which points lie on a given Cartesian equation's graph? Those whose coordinates satisfy the equation. Which points lie on a polar equation's graph? This requires a bit more care. By definition, a point lies on a polar equation's graph if ***at least one*** of its polar representations satisfies the equation. Consider point P in the figure. Does it lie on the graph of $r = \sin(\theta/2)$? The point's most obvious polar coordinates, $(1/2,\ 4\pi/3)$, do *not* satisfy the equation, so we might be tempted to conclude that P doesn't lie on the graph. This would be wrong; P has other polar representations, such as $(-1/2\,, 7\pi/3)$, which *do* satisfy the equation, so it does lie on the equation's graph.

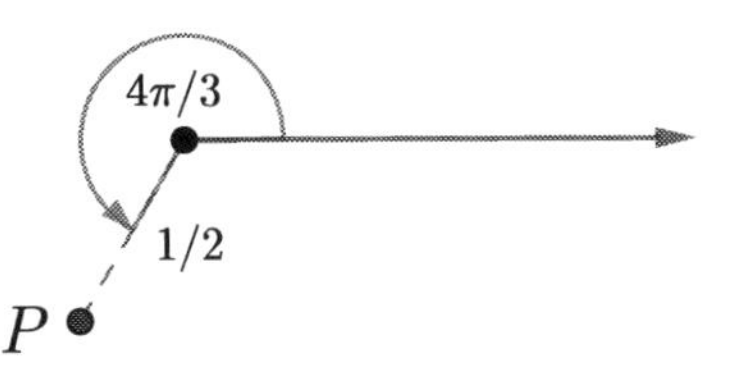

Polar equations, even quite simple ones, can have remarkably intricate graphs, as seen at right. Although computers produce such graphs with greater fidelity than we could ever hope to achieve *au natural*, graphing with our bare hands remains an invaluable, and perhaps indispensable, way to develop a feel for all things polar. This is especially true in the case of polar *functions*, which take the form $r = f(\theta)$.

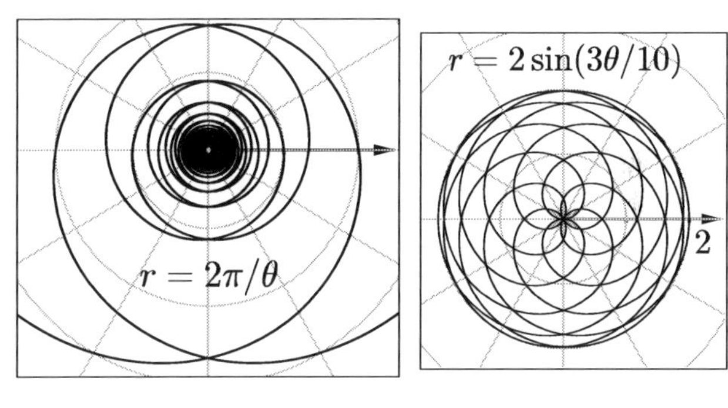

When graphing a polar function $r = f(\theta)$, it helps to think dynamically. Specifically, we'll imagine a ray rotating counterclockwise about the origin. We'll also imagine that a point – the tip of a pen – lies on the rotating ray and slides along it (towards or away from the origin) as the ray rotates, thus tracing out the polar graph. The "program" that tells the point how to move in response to the ray's rotation is the polar function itself, which specifies an r-coordinate for each angle θ through which the ray revolves.

For example, consider the polar function $r = f(\theta) = \theta/2\pi$. Since $f(0) = 0$, the point that traces out the equation's graph will start its journey at the origin. What happens to it when the ray begins rotating? Well, staring at the function and thinking about it will convince you that as the ray rotates (that is, as θ increases), r's value increases; hence, as the ray rotates, the point will slide ever further from the origin. After one full revolution of the ray, the point's distance from the origin will have increased from 0 to $f(2\pi) = 1$. After k revolutions, this distance will have become $f(2\pi k) = k$. The resulting spiral is shown at right.

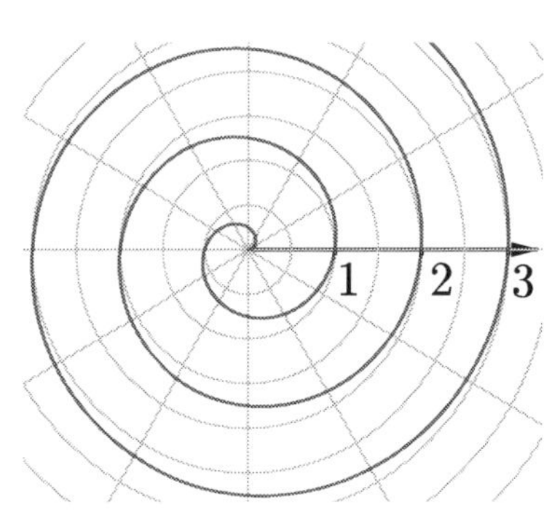

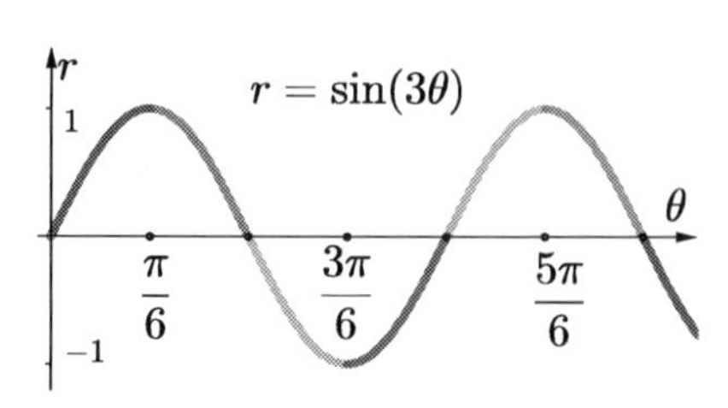

Graphing a polar function $r = f(\theta)$ as though it were Cartesian can help us sketch its polar graph. For example, consider $r = \sin(3\theta)$. By graphing this function on Cartesian axes, as at right, we produce an easy-to-read blueprint that will help us construct the genuine article, the polar graph. I've colored the Cartesian graph to help you see exactly how its parts corresponds to those on the polar graph we are about to produce.

To draw the polar graph of $r = \sin(3\theta)$, we consult our Cartesian blueprint and think dynamically. First, the blueprint tells us that when θ is zero, so is r. Hence, our polar graph will begin at the origin. Then, as the ray rotates, the Cartesian blueprint indicates how its mobile point slides along it in response. For instance, the blueprint's first portion (in black) indicates that as the ray's angle of rotation, θ, increases from zero to $\pi/6$, the point's distance from the origin, r, increase from zero to one. This observation allows us to draw the first (black) portion of the polar graph at right. Next, the blueprint's red portion reveals that as the ray rotates from $\pi/6$ to $2\pi/6$, the moving point returns to the origin. This lets us draw the red portion of the polar graph. Something peculiar happens in the green part of the blueprint: As the ray rotates from $2\pi/6$ to $3\pi/6$, we see that r's value becomes *negative*. As explained in the previous section, this means that the point drifts to the "wrong side" of the ray, appearing in the *third* quadrant, when we'd expect it, like the ray, to be in the first. Thinking your way through the entire analysis of the blueprint is excellent practice in understanding how this process works.

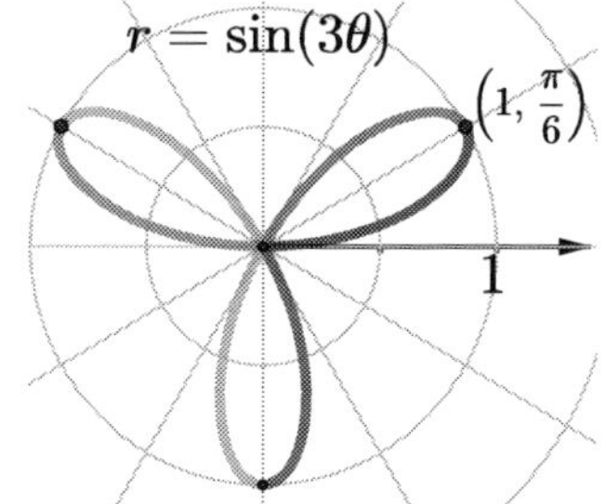

Exercises.

3. Sketch graphs of the following polar relations.

a) $r = 2$ b) $\theta = \pi/4$, with $r \geq 0$. c) $\theta = \pi/4$ with no restrictions on r. d) $0 \leq r < 1$ e) $0 < \theta < \pi/2$

4. Give polar relations that describe the following portions of the plane.

a) The polar axis (which is a *ray*) b) The line on which the polar axis lies

c) The circle of radius 3 centered at the origin d) The region between the unit circle and the circle in part (c).

5. Sketch graphs of the following polar functions. Give the coordinates of intersections with $\theta = 0$ and $\theta = \pi/2$.

a) $r = \theta/4\pi$, with $0 \leq \theta \leq 4\pi$. b) $r = \sin(2\theta)$ c) $r = 1 + \cos\theta$ d) $r = 1 - \cos(2\theta)$

e) $r = 1 - 2\sin\theta$

6. Can a point on the graph of $r = 1$ have a pair of polar coordinates for which $r \neq 1$? Explain.

7. Contemplate the graph of $r = \ln\theta$ at right until you can explain its basic features.

a) Is the figure at right the *complete* graph of the function, or is there more to it?

b) Why does the graph have a long "tail" sticking out to the left?

c) Part of the graph seems to be solid black. This might suggest that every point within that ring-shaped region lies on the graph. Is this the case?

d) As the graph winds around and around, why do its successive layers grow closer and closer together?

e) Use a graphing program to graph $r = \ln(\ln\theta)$. Its graph looks somewhat similar, but its tail points in another direction. Explain why.

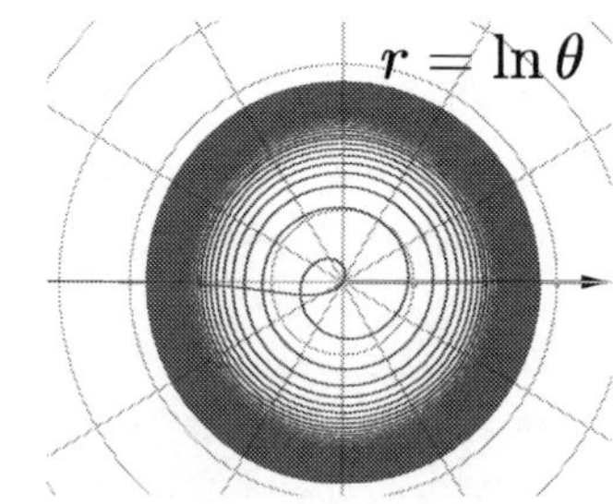

Conversions

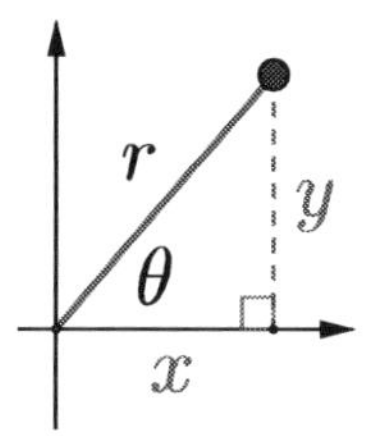

To convert between the Cartesian and polar coordinate systems, we identify the positive x-axis with the polar axis. A glance at the figure reveals the conversion formulas:

Polar to Cartesian: $\boldsymbol{x = r\cos\theta}$ and $\boldsymbol{y = r\sin\theta}$.

Cartesian to Polar: $\boldsymbol{r^2 = x^2 + y^2}$ and $\boldsymbol{\tan\theta = y/x}$.*

These formulas obviously hold for points in the first quadrant, but do they hold everywhere else in the plane? Indeed they do, as you'll verify in the exercises.

The conversion formulas can help us transform a Cartesian equation into polar form, or vice-versa. For example, graphing $r = 2\sec\theta$ with the previous section's methods might seem difficult. (Try it!) However, if you rewrite it as $r\cos\theta = 2$, then you've put it in a form in which the conversion formulas can help; they tell us that our equation is equivalent to $x = 2$, the graph of which is simply a vertical line.

Another example: Stop reading, and graph $r = 2\sin\theta$ with the previous section's methods. You'll end up with something like the shape at right: You'll be sure that it's a closed loop, but is it an ellipse, a circle, or something else? A clever observation will help us: Multiplying both sides of the equation by r turns it into $r^2 = 2r\sin\theta$, both sides of which should trigger our conversion reflexes; it is, in fact, equivalent to $x^2 + y^2 = 2y$. Completing the square turns this into $x^2 + (y-1)^2 = 1$, so the graph is a *circle*. (And should be redrawn accordingly.)

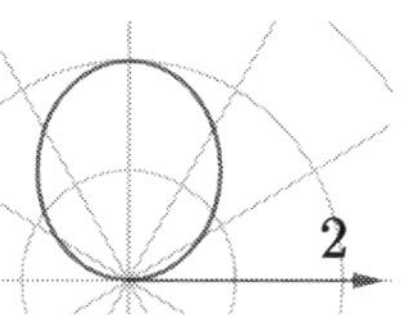

Exercises.

8. The polar coordinates of several points are given below. Give their Cartesian coordinates.

a) $(5, \pi)$ b) $(5, 210°)$ c) $(0,1)$ d) $(10, -\pi/4)$ e) $(-2, 360120°)$

9. Convert the following Cartesian coordinates to polar coordinates, with $r \geq 0$ and $0 \leq \theta < 2\pi$.

a) $(3,0)$ b) $(1, -\sqrt{3})$ c) $(-1,1)$ d) $(-\sqrt{3}, -1)$

10. We drew a picture above that shows why the conversion formulas hold for any point in the first quadrant. To understand why they hold for points in the *second* quadrant, consider the figure at right. If the point's Cartesian and polar coordinates are given by (x, y) and (r, θ) respectively, explain why the conversion formulas hold. Once you've done that, draw the relevant figures for points in the third and fourth quadrants and explain why the conversion formula hold there as well.

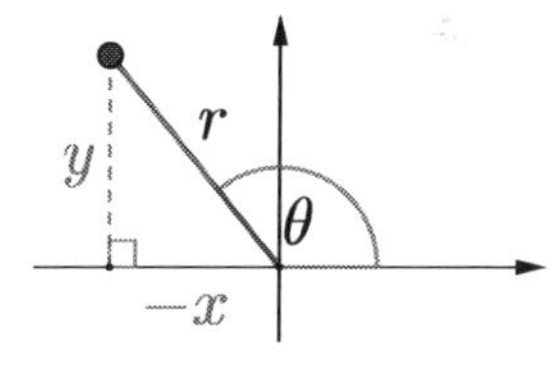

11. The conversion formulas hold for points on the horizontal axis, too. Explain why. For points on the vertical axis, all but one of the conversion formulas hold. Point out the one that fails in this case.

12. Convert the following Cartesian equations to polar equations.

a) $x^2 + y^2 = 4$ b) $y = 2$ c) $2x + 3y = 5$ d) $(x^2 + y^2)^2 = 9(x^2 - y^2)$ e) $y = x^3 + xy^2$

13. Convert the following polar equations to Cartesian equations.

a) $r\cos\theta = -1$ b) $r = 3\csc\theta$ c) $r = 5\cos\theta$ d) $r = -2$ e) $r = 2/(1 + \cos\theta)$ f) $\theta = \pi/6$

---

* Even down here in the footnote basement, I heard your cry: "Why doesn't he just write $\theta = \arctan(y/x)$?" Recall, young Padawan, that arctangent's range is $(-\pi/2,\ \pi/2)$, which means that if the point whose coordinates you are converting happens to lie in the second or third quadrant, the arctangent formula you've proposed will *not* produce the correct value.

Areas in Polar Coordinates

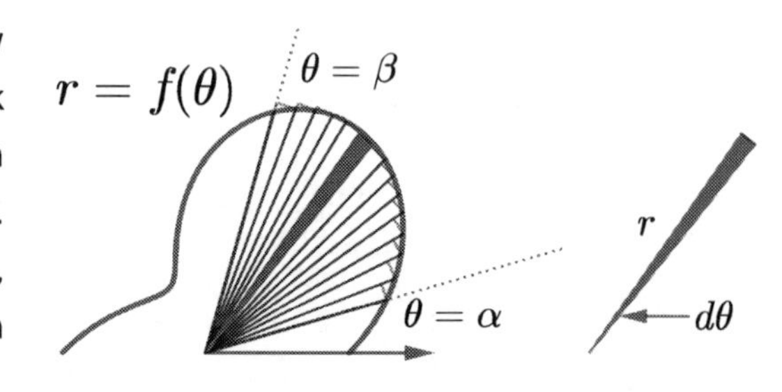

As a veteran of the calculus wars, you already know how to find the area under a Cartesian function's graph: Break the region into a set of infinitely many infinitesimally thin rectangles, then sum all of their areas up with an integral. To find the area swept out by a *polar* function, however, it's better to imagine the region as an infinite collection of *slices of pie*, each of which is infinitesimally thin.*

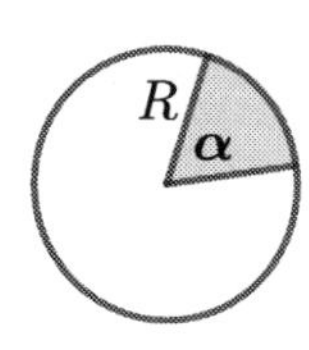

Recall that a pie slice of radius R and central angle α, as in the figure at right, has area $(1/2)R^2\alpha$. [Proof: Since its central angle is $\alpha/2\pi$ of a complete rotation, the slice takes up that fraction of the circle's complete area. Thus, its area is $(\alpha/2\pi)\pi R^2 = (1/2)R^2\alpha$.] Applying this to our problem above, we observe that a typical slice of the fanlike region whose area we want has radius r and central angle $d\theta$, so its area must be $(1/2)r^2 d\theta$. By integrating this expression over all relevant values of θ, we obtain an expression for the region's area:

$$\int_\alpha^\beta \frac{1}{2} r^2\, d\theta.$$

Don't just memorize this expression; understand it. To verify that it performs as advertised, we'll first use it to compute something whose value we already know – a circle's area.

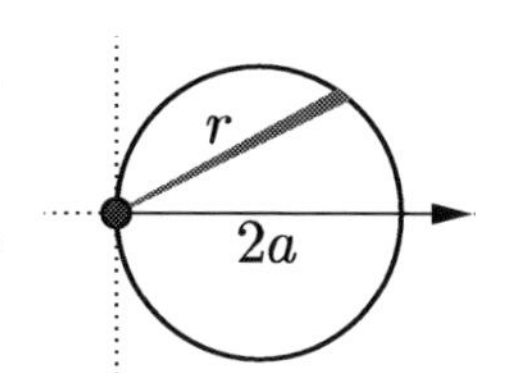

Example 1. The graph of $r = 2a\cos\theta$ is a circle of radius a, as shown at right. Confirm that its area is πr^2 by computing an integral.

Solution. Draw a typical slice and note its area: $(1/2)(2a\cos\theta)^2 d\theta$. To simplify our work, we'll find the area of the circle's top half and double the result:

$$2\int_0^{\pi/2} (1/2)(2a\cos\theta)^2 d\theta.$$

To understand the integral's boundaries, you must consider how the curve is traced out dynamically. When θ is 0, the moving point is $2a$ units down the polar axis; as θ increases to $\pi/2$, it traces out the top half of the circle counterclockwise. Working the integral out (do it: it is good practice), you'll find that its value is indeed πa^2. ♦

When finding areas this way, you must take care when determining the boundaries of integration.

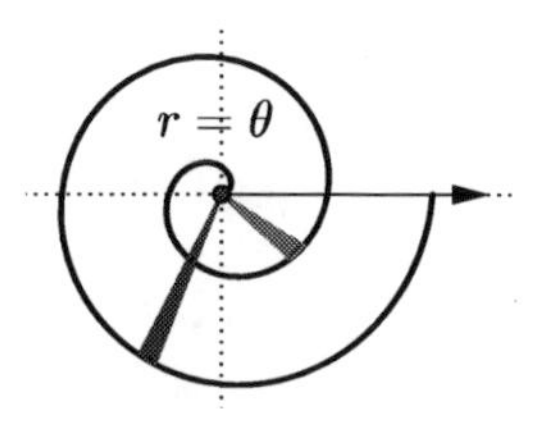

Example 2. Find an expression for the area contained in the "snail shell" at right.

Solution. The figure shows the portion of the graph of $r = \theta$ for which θ is between 0 and 4π. A typical pie slice (like the short one at right) will have area $(1/2)r^2 d\theta = (1/2)\theta^2 d\theta$. However, integrating this expression from 0 to 4π would yield the *wrong* value for the snail shell's area. To understand why, you must think carefully about what happens geometrically when we integrate.

* "It's only wafer thin…" (Apologies to deprived souls who know not Monty Python's *Meaning of Life*.)

Namely, As θ runs from 0 to 2π, our pie slices sweep up all the area in the shell's first revolution. However, once θ passes 2π, the subsequent pie slices (such as the *long* one in the preceding figure) contain some "new" area in the shell's second revolution, but they also contain some "old" area within the shell's first revolution – area that we've already summed up in our integral. Hence, we'd be adding all that stuff *twice* if we were to integrate pie slices from 0 to 4π.

Having recognized the problem, we can now think of various ways to get around it if we wish to find the shell's area. The simplest expression for this area would be

$$\int_{2\pi}^{4\pi} \frac{1}{2}\theta^2 d\theta.$$

Be sure you understand geometrically why this integral will yield the correct area. (Think of the various pie slices that get summed up as θ runs from 2π to 4π.) ♦

Here's one last example.

Example 3. In the figure at right, find the area of the crescent-shaped region lying inside the circle but outside the heart.

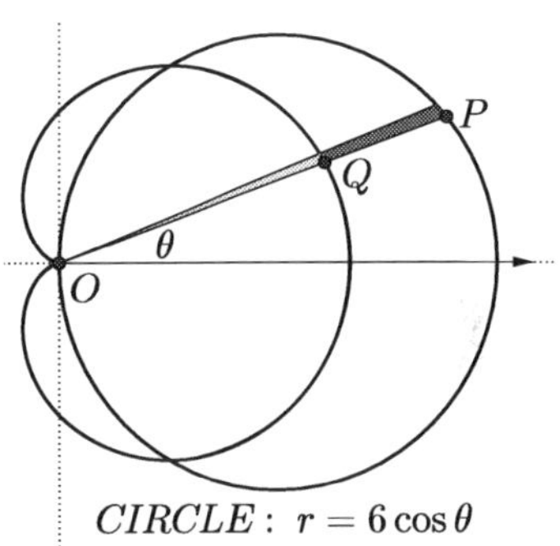

Solution. For any given value of θ, we want the area of the dark part of the pie slice (from Q to P); since this dark part is the difference of *two* full pie slices (one from O to P, one from O to Q), its area is $(1/2)(6\cos\theta)^2 d\theta - (1/2)(2(1+\cos\theta))^2 d\theta$, which reduces to $\boldsymbol{(16\cos^2\theta - 4\cos\theta - 2)d\theta}$, as you should verify. To find the crescent's area, we must integrate this expression between those values of θ at which the circle crosses the heart (and hopes to die).

To find these intersection points, we equate the expressions for r in the circle and heart equations. Simplifying the result yields $\cos\theta = 1/2$; since this occurs when $\theta = \pm\pi/3$, we could find the crescent's area by integrating the expression that we found above from $-\pi/3$ to $\pi/3$. But, to make life a bit easier, we'll exploit the figure's symmetry and compute the following integral:

$$2\int_0^{\pi/3} (16\cos^2\theta - 4\cos\theta - 2)d\theta.$$

If you work this out, you'll find that the crescent's area is exactly 4π. ♦

One last time: When computing areas in polar coordinates, you must take pains to think your way through each problem to ensure that the integral you set up actually represents the area you want.

Exercises.

14. Sketch the graph of $r = \sqrt{\sin\theta}$, and find the area contained within it. When you are finished, graph the function on a computer to check your graph's accuracy.

15. a) If $r = f(\theta)$ is an *even* polar function, its graph exhibits a particular type of symmetry. Which type and *why*?
b) Same story for an *odd* polar function.

16. Find the area contained within the heart $r = 2(1 + \cos\theta)$, whose graph is shown in Example 3 above.

17. Using nothing but classical geometry, Archimedes proved (in his treatise *On Spirals*) a result that we'd state as follows: The region contained in one revolution of the spiral $r = \theta$ is exactly 1/3 of the area contained in the circle centered at the origin and passing through the endpoint of the spiral's first revolution.

Your problem: Sketch a picture of the spiral and circle, then use calculus to prove that Archimedes is correct.

18. Find the area of the region that lies within both of these circles: $r = \sin\theta$ and $r = \sqrt{3}\cos\theta$.

19. The graph of $r^2 = 2\cos(2\theta)$ is called a *lemniscate*. Find the area that it encloses.
[*Hint: Be careful with those boundaries of integration! You'll need to think about the equation, not just its graph, to determine what they should be.*]

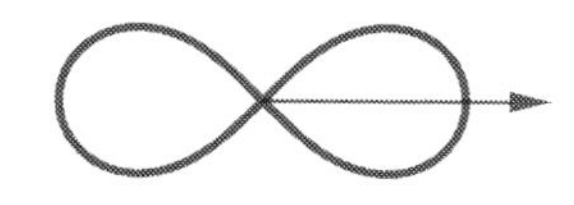

20. a) Make a rough sketch of the graph of $r = \tan(\theta/2)$ and find the area it encloses.
b) A more careful sketch would indicate the long-run behavior of the graph's "arms". What happens to the arms as they stretch away from the graph's central loop?

21. Sketch the graphs of $r = \sin 2\theta$ and $r = \cos 2\theta$ and find the area of the region lying within both curves.

22. In Example 3, we needed to find the intersections of the circle $r = 6\cos\theta$ and the heart-shaped graph of $r = 2(1 + \cos\theta)$. We did this by setting their right-hand sides equal to one another. This yielded an equation whose solutions were $\theta = \pm\pi/3$. Because that information sufficed to solve the larger problem posed in Example 3, you may have been too busy celebrating to have noticed that a *third* intersection (at the origin) somehow went undetected by our algebraic process.

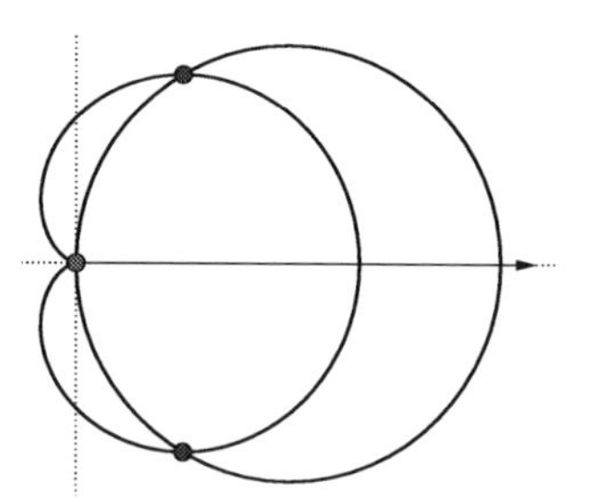

Question: Why did our algebra fail to identify the intersection at the origin?

23. (Optional – for insatiable delvers into infinitesimal details.)

We've been using "pie slices" to find areas, but why does this technique actually work? Return to basic principles (as in Chapter 6, exercise 12) and recall that curves are locally *straight*. This means that a typical slice isn't actually a pie slice (i.e. with a circular edge), but rather an infinitesimally thin *triangle*, whose infinitesimal (and straight!) side coincides with the (locally straight) curve. As always, an infinitesimal change $d\theta$ in the function's input begets a corresponding infinitesimal change dr in its output, so if our triangle's initial side has length r, its terminal side will have length $r + dr$.

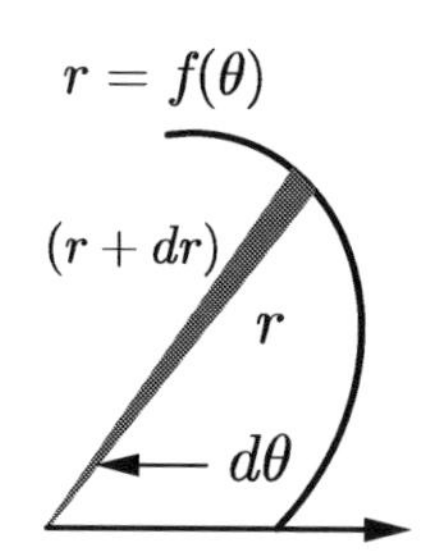

a) Explain why the area of the infinitesimally thin triangle is $\frac{1}{2}\,r(r + dr)\sin(d\theta)$.

b) Substitute $d\theta$ into the Maclaurin series for sine, recalling from Chapter 1 that products of infinitesimals may be treated as zeros, and conclude that $\sin(d\theta) = d\theta$.

c) Combining the results of parts (a) and (b), explain why our infinitesimal triangle's area is $\frac{1}{2}r^2 d\theta$.

d) Explain why this justifies the validity of our "pie slice" technique.

Slopes in Polar Coordinates

Any "nice" curve is locally straight, so we can speak of its *local* slope at a given point, or in other words, its "infinitesimal rise over an infinitesimal run" (which we denote dy/dx) at that point. If the curve is the graph of some function $y = f(x)$, then computing dy/dx at a given point is a typical Calculus 1 problem. But what if the curve is the graph of a *polar* function $r = f(\theta)$, in which x and y are conspicuously absent?

Thanks to the uncanny power of Leibniz notation, finding a polar graph's slope at a given point is easy. The key is to observe that the ordinary rules for fractions give us

$$\frac{dy}{dx} = \frac{\frac{dy}{d\theta}}{\frac{dx}{d\theta}}.$$

Coupling this with our conversion formulas ($y = r\sin\theta\,,\ x = r\cos\theta$), we can find a polar curve's slope.

Example. Find the slope of the graph of $r = 1 + 5\cos\theta$ at the point $\left(1 + \frac{5}{\sqrt{2}},\ \frac{\pi}{4}\right)$.

Solution. Observe that for any typical point on the graph, we have

$$\frac{dy}{dx} = \frac{\frac{dy}{d\theta}}{\frac{dx}{d\theta}} = \frac{\frac{d}{d\theta}(r\sin\theta)}{\frac{d}{d\theta}(r\cos\theta)} = \frac{\frac{d}{d\theta}\big((1+5\cos\theta)\sin\theta\big)}{\frac{d}{d\theta}\big((1+5\cos\theta)\cos\theta\big)}.$$

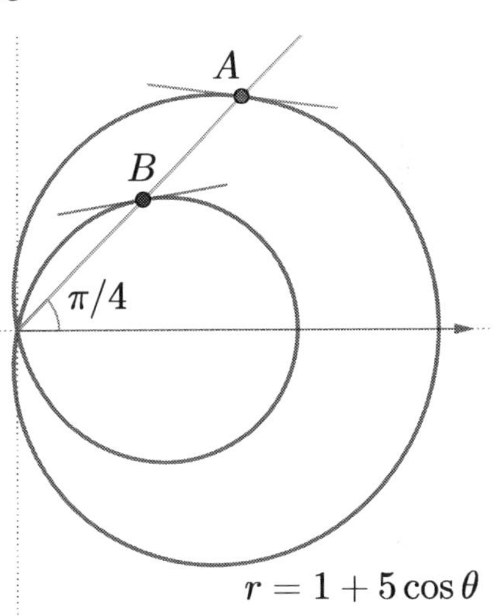

The rest is just computation. Grinding out the derivatives yields

$$\frac{dy}{dx} = \frac{\cos\theta + 5\cos(2\theta)}{-(\sin\theta + 5\sin(2\theta))},$$

as you should verify. It follows that at the point we're considering (point A on the graph), the curve's slope must be

$$\frac{\cos\frac{\pi}{4} + 5\cos\frac{\pi}{2}}{-\left(\sin\frac{\pi}{4} + 5\sin\frac{\pi}{2}\right)} = \cdots = \frac{-\sqrt{2}}{\sqrt{2}+10} \approx -0.124\,. \quad \blacklozenge$$

That's all there is to it. And yet... nothing involving polar coordinates should be done on autopilot; those multiple polar representations of points can confuse us terribly if we forget about them. For example, in the preceding example, our explicit formula for dy/dx in terms of θ would seem to imply that since points A and B both lie on the line $\theta = \pi/4$, the curve's slope should be equal at both points. However, a glance at the figure shows that this clearly is not the case. Why not? The answer is both simple and subtle.

Like all points, B has many polar representations. In some, θ is in the first quadrant and r is positive; in others, θ is in the third quadrant and r is *negative*. It can be shown, interestingly, that only the latter representations satisfy this particular curve's equation. Consequently, to the eyeless god of algebra (for whom a "curve" is not a geometric object, but a set of ordered pairs of numbers), the point we call B does not have a θ-coordinate of $\pi/4$. Rather, B is on the curve because it can be represented by coordinates that *do* satisfy the curve's equation, such as $\left(1-(5/\sqrt{2}), \mathbf{5\pi/4}\right)$. Thus, to find the curve's slope at B, we could substitute $5\pi/4$ into our dy/dx formula, obtaining a slope of approximately 0.165.

Determining slopes and areas associated with polar curves isn't (let us be honest) a particularly pressing scientific problem. As with so much else in mathematics – nay, in life itself – the value in studying these things lies not in the things themselves. Rather, the point is for you to develop facility with certain modes of thought. The real goals here are for you to extend your experience with the somewhat slippery system of polar coordinates, to understand how polar coordinates relate to Cartesian coordinates, and to appreciate the power of infinitesimal thinking and Leibniz's ingenious notation. These are vital lessons for anyone learning to assimilate mathematics into his or her mental toolkit. Solving slope and area problems by *thinking* your way through them deepens your understanding. Solving them by memorizing and applying ill-understood algorithms gleaned from the internet deepens nothing. Be sure to avoid this trap.

Exercises.

24. Find the slope of the spiral $r = 5/\theta$ at $\left(\frac{15}{\pi}, \frac{\pi}{3}\right)$.

25. Find the slope of the flower $r = \sin(5\theta)$ at $\left(-\frac{\sqrt{3}}{2}, \frac{\pi}{3}\right)$.

26. The graph of $r = \sin(\theta/2)$ is shown at right.

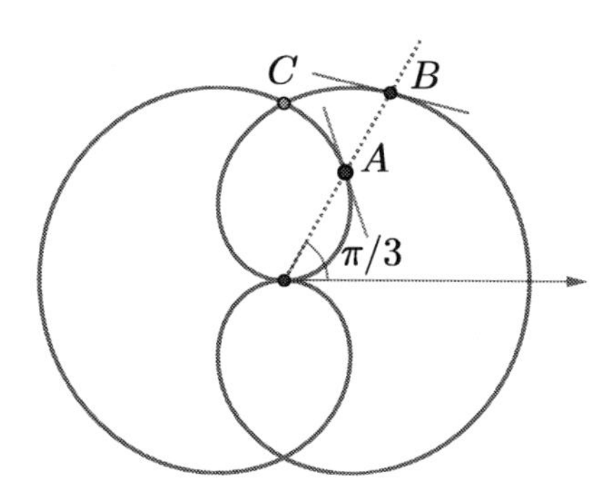

a) Find polar coordinates for A and B that satisfy the curve's equation. [*Hint: Substitute different polar representations of points A and B into the curve's equation. Some will satisfy it. Others won't.*]

b) For each point (A and B), find a pair of polar coordinates that does *not* satisfy the curve's equation.

c) Find the curve's slope at A and at B.

d) Unlike graphs of Cartesian functions, graphs of polar functions can intersect themselves. This one, for example, crosses itself at C. Convince yourself that in an infinitesimal neighborhood of C, the graph is the intersection of two straight lines.

e) Find the slopes of the two lines mentioned in part (d).

f) The graph of $r = 1 + 5\cos\theta$ (its graph is shown in the example above) has a self-intersection at the origin, where it – locally – resembles two straight lines. Find the slopes of these lines.

27. The authors of some calculus texts, paying no heed to the thoughts with which I ended this section, offer their readers the following recipe for finding the slope of a polar function $r = f(\theta)$ at a point (r, θ):

$$\frac{\frac{dr}{d\theta}\sin\theta \;+\; r\cos\theta}{\frac{dr}{d\theta}\cos\theta \;-\; r\sin\theta}.$$

Explain why this recipe does indeed give the desired slope. (And, for your own sake, please don't memorize it.)

28. (Who needs hallucinogens when you've got polar coordinates?) If you are able, instruct a computer to graph the polar equation $r = a\cos^2(b\theta) + c\sin^2(d\theta)$ while the values of the four parameters $a, b, c,$ and d vary – preferably at different speeds. [*Hint: You **are** able to do this. At the time I'm writing this, the free online graphing calculator Desmos handles this sort of thing easily. Just type in the equation and follow your nose from there.*]

Your assignment: Enjoy – and try to find interesting variations that will produce still wilder polar graphs.

Parametric Equations

"Eppur si muove."

- Galileo Galilei

Imagine a point moving in the coordinate plane, tracing out a curve as it moves. If at time t, the point's coordinates are $(f(t), g(t))$, we say that $x = f(t)$ and $y = g(t)$ are **parametric equations** for the curve.

Example. Sketch the graph given by the parametric equations $x = t^3$ and $y = t^2$.

Solution. By considering specific t-values and the x's and y's corresponding to them, we discern the moving point's location at various times. For example, when $t = 1$, the point is at (1,1), whereas when $t = 2$, it is at (8,4). Mentally calculating a few more locations of the moving point reveals its trajectory: As t increases from zero, x and y increase, but x grows faster, since it *cubes* t's value, while y merely squares it. As t *decreases* from zero, the point moves leftwards and upwards from the origin. ♦

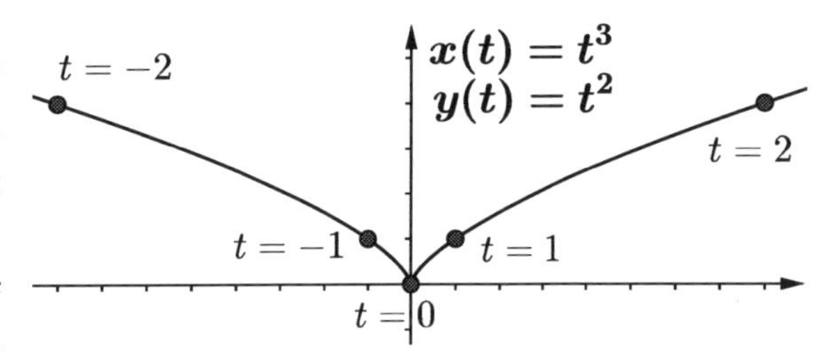

The independent variable t on which the moving point's coordinates depend is called a *parameter*. Any symbol can serve as a parameter, although t is the most common, probably because it suggests "time". Sometimes, however, a parameter has a specific *geometric* meaning, as in the best-known of all parametric equations – those for the unit circle, which, as all trigonometry students learn, are $x = \cos\theta$ and $y = \sin\theta$, where the parameter θ is the angle in the figure at right.

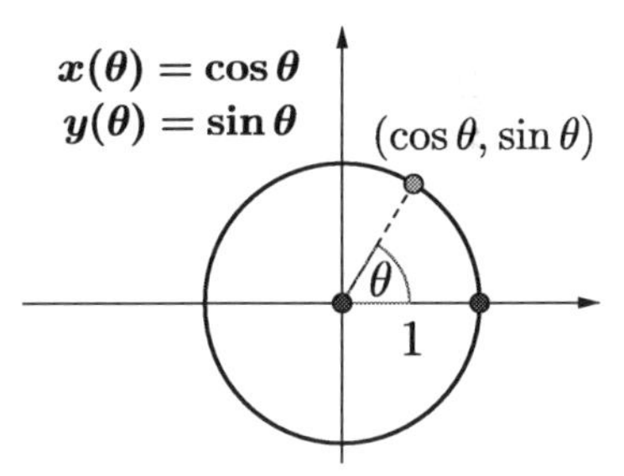

Given a set of parametric equations, we can sometimes eliminate the parameter and combine them into a single Cartesian equation. For example, given $x = t^3$, $y = t^2$ from the example above, we can solve the first equation for t and substitute the result into the second. This yields $y = x^{2/3}$, which is the graph's Cartesian equation. Similarly, squaring and adding the parametric equations $x = \cos\theta$ and $y = \sin\theta$ yields $x^2 + y^2 = 1$, the unit circle's familiar Cartesian equation.

Because they facilitate the analysis of motion one component at a time, parametric equations are ubiquitous in physics. They are also common in geometry. Curves in 3-dimensional space, for example, are nearly always described parametrically; to parametrize a curve in space (as opposed to the plane), we need only add a third equation for the moving point's z-coordinate. For example, the *helix* shown at right is parametrized by

$$x = \cos t, \quad y = \sin t, \quad z = t.$$

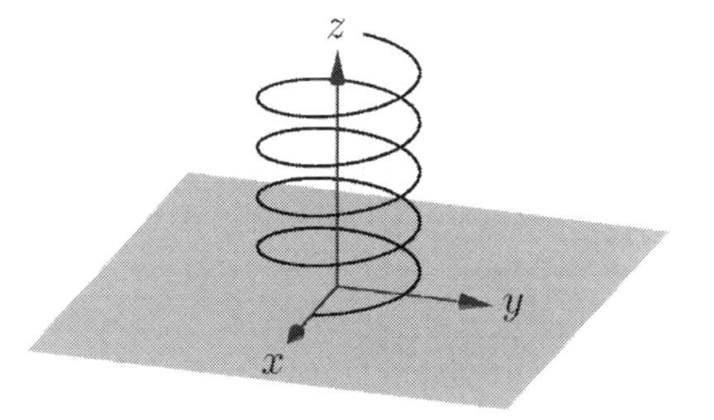

Indeed, nothing - apart from our imagination - limits us to three dimensions. With a set of n parametric equations, we can describe a curve in n-dimensional space. We can't see such a curve, but we can still study it analytically.

For our next example, we turn to the *cycloid*, a curve whose beautiful properties were the subject of intense study (and a certain amount of quarreling) among several giants of 17th-century mathematics. It isn't clear who first described the cycloid, but the man who named it was Galileo Galilei. By definition, the cycloid is *the curve that is traced out by a point on a circle as the circle rolls along a line*. The figure at right shows one arch of the cycloid, generated as the circle completes one full revolution while rolling on the line. If the circle were to continue rolling, the arch would, of course, be repeated over and over.

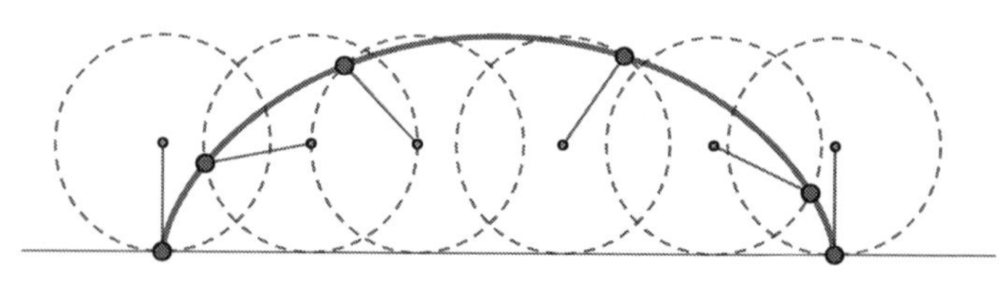

To parametrize the cycloid, assume that a circle of radius a rolls on the x-axis, and that the marked point P tracing out the cycloid begins at the origin. You'll need to ponder the figure for a while to grasp its various parts. The key idea is that if we were to "roll back" the circle to its initial position, then radius CP would rotate counterclockwise through an angle of θ until it ended up lying along the y-axis, and point P would be returned to its initial position at the origin. We shall use angle θ, which measures the angle through which the radius CP has turned as the circle rolls, as our parameter. With this in mind, we can now derive the cycloid's parametric equations.

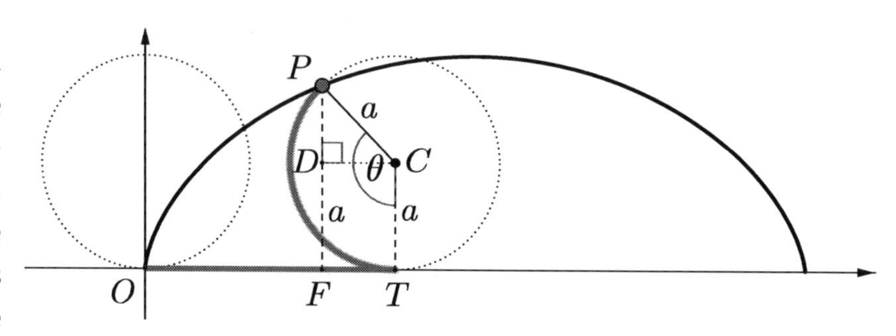

If we call P's coordinates x and y, the figure above will guide us to the cycloid's parametric equations:

$$\boldsymbol{x} = OF = OT - FT = \text{arc}(TP) - DC = a\theta - a\cos\left(\theta - \frac{\pi}{2}\right) = a\theta - a\sin\theta = \boldsymbol{a(\theta - \sin\theta)},$$

$$\boldsymbol{y} = FP = FD + DP = a + a\sin\left(\theta - \frac{\pi}{2}\right) = a - a\cos\theta = \boldsymbol{a(1 - \cos\theta)}.$$

Be sure that you can justify all those equals signs!

Historical cycloid stories could fill an entire book. I'll mention just two here. First, while investigating the principles of the ideal pendulum clock, Christiaan Huygens solved the so-called *tautochrone* problem, which seeks a curve with the following property: A marble, if released and allowed to slide down this curve under the force of gravity, should take the same amount of time to reach the bottom *regardless of the position on the curve from which it was released*. Huygens discovered (and, in 1673, published the result) that there is indeed a tautochrone curve: namely, the cycloid, turned upside down.* Second, although Pascal had abandoned mathematics for theology in 1654, when a terrible toothache kept him awake one night in 1658, he began thinking about the cycloid in the hopes that intense thought would distract him from intense pain. It worked: the pain subsided entirely, which Pascal took as a divine sign. He proved several theorems about cycloids, published the solutions, and then promptly re-abandoned mathematics.

* In Chapter 96 of *Moby Dick*, Ishmael polishes the interior of the Pequod's try-pots (in which oil was extracted from blubber), noting as he does so, "It is a place also for profound mathematical meditation... with the soapstone diligently circling round me... I was first indirectly struck by the remarkable fact, that in geometry all bodies gliding along the cycloid, my soapstone for example, will descend from any point in precisely the same time."

Exercises.

29. Sketch the graphs of the following parametric curves by any means necessary. [*Hints: Plot specific points, think more generally about how the moving point varies its position as the parameter t increases from large negative values to zero and from zero to large positive values, eliminate the parameter to convert to a Cartesian equation. Try to understand each graph's shape in several different ways.*]

a) $x = 2t, \quad y = 1 + t$ b) $x = t^2, \quad y = 2t$ c) $x = 2^t, \quad y = 2^{t+1}$

30. a) The graph of any function $y = f(x)$ can be parametrized very easily. Explain how.

b) Find parametric equations for the graph of $y = \sin(1/x)$.

c) Find parametric equations for the line through the origin that makes an acute angle θ with the positive x-axis.

31. A given curve's parametric equations are *not* unique; it can be parametrized in multiple ways. To grasp this important point, note that each of the following five sets of parametric equations has the same graph: the line $y = x$. In each case, however, the moving point traces the line out in a different manner.

Your problem: Compare and contrast the motions of the moving point in each case.

a) $x = t, \quad y = t$ b) $x = t^3, \quad y = t^3$ c) $x = 2 + 3t, \quad y = 2 + 3t$

d) $x = 1 - \sqrt{2}t, \quad y = 1 - \sqrt{2}t$ e) $x = t\sin t, \quad y = t\sin t$

32. The so-called *Witch of Maria Agnesi* is a curve generated as follows:

Center a circle of radius 1 at $(0, 1)$.
Let a ray from the origin sweep from $\theta = 0$ to π.
Let A and B be the ray's intersections with $y = 2$ and the circle.
Let P be the point with A's x-coordinate and B's y-coordinate.
The curve traced out by point P as the ray rotates is the witch.

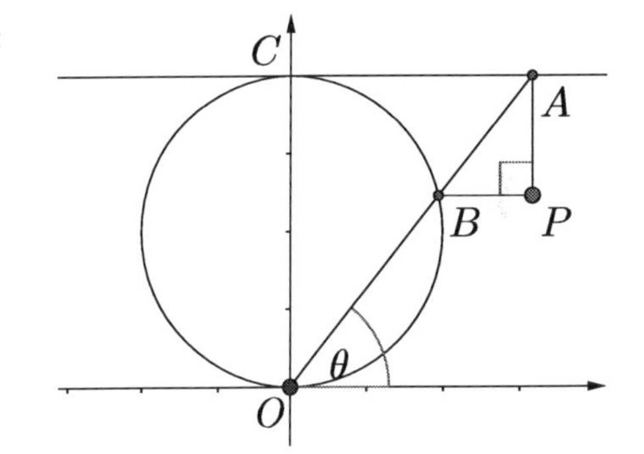

a) Draw the witch.

b) Find parametric equations for it, using angle θ as the parameter.

c) Learn the source of the strange name.

33. For a particularly baroque example of a parametric curve, type "Cupid curve" into Wolfram Alpha. How the parametric equations for the Cupid curve were generated is a mystery to me. Perhaps you can find out.

34. As of this writing, Geogebra has a command ("curve") for plotting parametric curves, but lacks a command for plotting polar functions. One can, however, plot a polar function $r = f(\theta)$ indirectly – by parametrizing it! Explain why the graph of the parametric equations

$$x(t) = f(t)\cos t, \qquad y(t) = f(t)\sin t$$

is identical to the graph of the polar function $r = f(\theta)$. Then use this trick to parametrize the polar equation $r = \sin(\theta/2)$, and graph the result on Geogebra. (You can check your graph by comparing exercise 26 above.)

35. By considering the moving point's motions, describe the following parametric equations' graphs:

a) $x = 2 + 3t, \quad y = 4 - 2t$ b) $x = 2 + 3t, \quad y = 4 - 2t, \quad z = 5 + t$

c) $x = t\cos t, \quad y = t\sin t$ d) $x = t\cos t, \quad y = t\sin t, \quad z = t$ (where $t \geq 0$).

36. Regarding the figure at right, a cycloid generated by a circle of radius a,

a) How long is the base OQ of the arch?

b) At points O and Q, what are the values of the parameter, θ?
[*Hint: Review the figure on the previous page in which θ is defined.*]

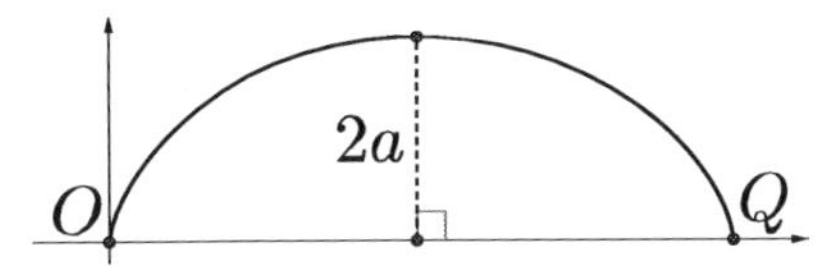

Calculus and Parametric Equations

As might be expected, Leibniz' brilliant notation lets us apply calculus to parametric equations with ease. To see how, let's ask a couple of calculus-style questions about the cycloid.

Problem 1. What is the area under one arch of a cycloid?

Solution. Consider a cycloid generated by a circle of radius a. In the previous section, we derived its parametric equations:

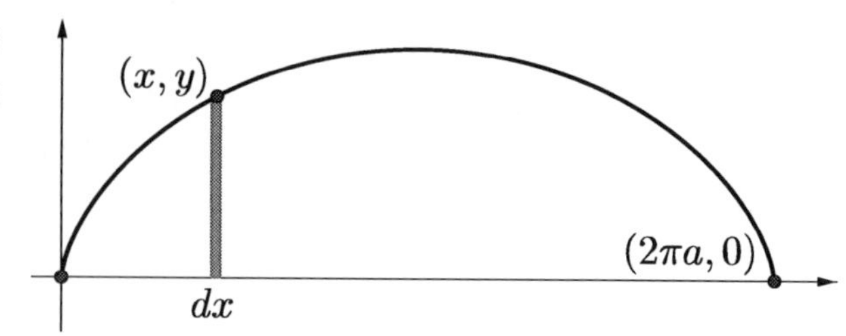

$$x = a(\theta - \sin\theta)$$
$$y = a(1 - \cos\theta).$$

If we imagine the region below the cycloid as consisting of infinitely many infinitesimally thin rectangles, then a typical rectangle standing at x will have – as should be clear from the figure – an area of ydx. From this, it follows that the total area under the cycloid must be

$$\int_0^{2\pi a} y\,dx,$$

which looks simple... until we remember that we don't know how to express y as a function of x.

Leibniz notation, however, helps us convert everything in the integral into the language of θ. The key is to observe that $dx = (dx/d\theta)d\theta$. This lets us rewrite the integral as

$$\int_0^{2\pi} a(1 - \cos\theta)\frac{dx}{d\theta}d\theta = \int_0^{2\pi} a(1 - \cos\theta)\,a(1 - \cos\theta)d\theta.^*$$

If you work out the integral, which you should, you'll find that its value is $3\pi a^2$. That is:

The area under a cycloidal arch is exactly 3 *times that of the cycloid's generating circle.* ♦

The area under a cycloidal arch was found by Evangelista Torricelli in 1644, when Newton was an infant and Leibniz but a gleam in his father's eye.† Torricelli discovered the area beneath a cycloid by means of his own "method of indivisibles", a forerunner of integral calculus. Calculus students should hold Torricelli in high honor for discovering "Gabriel's Horn" (also called "Torricelli's Trumpet"), whose paradoxical-sounding properties, which you explored in Chapter 6, exercise 16, greatly disturbed the leading mathematicians of his time, and continue to disturb the leading calculus students of our own time.

Another obvious geometric question about a cycloidal arch: How long is it? The exact answer was first discovered in 1658 by Christopher Wren, best known as one of England's greatest architects. Wren's discovery of the cycloid's length in the BC era (Before Calculus) testifies to his mathematical talent.

---

* To convert the integration boundaries to θ geometrically, recall exercise 36b. Alternatively, you could carry out the conversion analytically by noting that the cycloid's x-equation, $x = a(\theta - \sin\theta)$, tells us that when $x = 2\pi a$, we have $2\pi = \theta - \sin\theta$, which is obviously satisfied when $\theta = 2\pi$. (Similarly, when $x = 0$, we have $\theta = 0$.)

† Torricelli also invented the barometer in 1644, remarking poetically that "We live submerged at the bottom of an ocean of air." He died three years later at the age of 39.

Problem 2. What is the length of one arch of a cycloid?

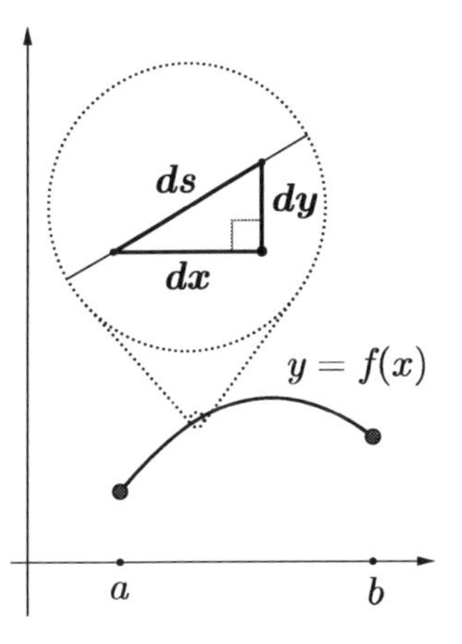

Solution. Recall the "differential triangle" at right (from Chapter 6), which lies at the heart of all arclength calculations. Since the cycloid is given by parametric equations in θ, we'd like to write ds in terms of θ and then integrate as θ runs from 0 to 2π. To get $d\theta$ into the picture, we'll divide the triangle's sides by $d\theta$, thus producing a similar triangle whose sides have lengths $ds/d\theta, dx/d\theta$, and $dy/d\theta$. Applying the Pythagorean Theorem to *that* triangle and solving for ds yields, as you should verify,

$$ds = \sqrt{\left(\frac{dx}{d\theta}\right)^2 + \left(\frac{dy}{d\theta}\right)^2}\, d\theta\,.$$

Substituting into this the cycloid's parametric equations, which, as we've seen, are

$$x = a(\theta - \sin\theta) \quad \text{and} \quad y = a(1 - \cos\theta),$$

yields an expression for the cycloid's infinitesimal element of arclength:

$$ds = \sqrt{(a(1-\cos\theta))^2 + (a\sin\theta)^2 d\theta} = \sqrt{2}a\sqrt{1-\cos\theta}\, d\theta.$$

Thus, the cycloid's total length is

$$\sqrt{2}a\int_0^{2\pi}\sqrt{1-\cos\theta}\, d\theta.$$

This is a tricky integral, but we can crack it with our trusty half-angle identity: Since

$$\frac{1-\cos\theta}{2} = \sin^2\left(\frac{\theta}{2}\right),^*$$

it follows that $1 - \cos\theta = 2\sin^2(\theta/2)$. Substituting this for the expression under the radical in the integral above, we find that the cycloid's length is

$$\sqrt{2}a\int_0^{2\pi}\sqrt{1-\cos\theta}\, d\theta = \sqrt{2}a\int_0^{2\pi}\sqrt{2\sin^2(\theta/2)}\, d\theta = 2a\int_0^{2\pi}\sin(\theta/2)\, d\theta = \cdots = 8a.$$

That is, *the cycloid's arclength is exactly* 8 *times its generating circle's radius.* ♦

In 1658, only an inspired thinker could have proved this result. Today, ordinary college freshmen can prove it, thanks to the fully-developed calculus of infinitesimals, which reduces the problem to a routine calculation – precisely as a calculus should. Newton himself famously wrote to Robert Hooke, "If I have seen further it is by standing on the shoulders of Giants." As this calculus book - this calculus course - comes to an end, and you dismount your giant, be sure to tip your hat to Wren, but bow especially deeply to Newton, Leibniz, and the others who created the magnificent machine you've been learning to operate.

---

* If this looks unfamiliar, start with the more familiar identity $\sin^2 x = (1 - \cos 2x)/2$, and substitute $\theta/2$ for x.

Exercises.

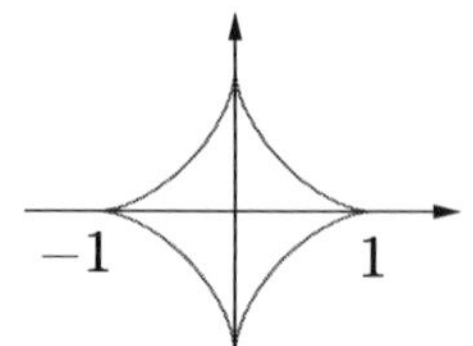

37. The *asteroid* shown at right is given by the parametric equations

$$x = \cos^3 t, \quad y = \sin^3 t.$$

a) Find the smallest value of t corresponding to each of the asteroid's four cusps.

b) Suppose you wish to find the asteroid's length. By the curve's symmetry, it suffices to analyze the part of it lying in the first quadrant, whose length is exactly one quarter of the full length. With this in mind, find ds, the arclength element for the asteroid in the first quadrant.

c) Using ds from the previous part, integrate to compute the length of the portion of the curve that lies in the first quadrant. Then quadruple the result to obtain the length of the full asteroid.

d) Without actually integrating, decide whether the following expression is equal to the asteroid's total length:

$$2\int_0^\pi 3\cos t \sin t \, dt.$$

After answering, compute the integral. The result may surprise you. If so, explain why this happened.

38. Find an integral representation of the area enclosed by the asteroid in the previous problem. Use a computer to carry out the actual integration. (It is possible – but unpleasant – to do this integral by hand.)

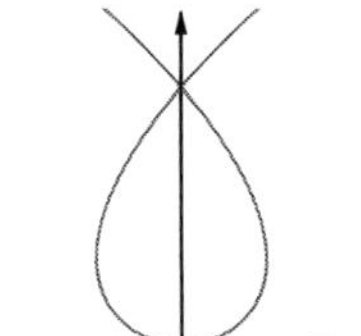

39. The curve shown at right is given by the parametric equations

$$x = t^3 - 3t, \quad y = 4t^2.$$

a) Find the coordinates of the point at which the curve crosses itself.

b) Find the area enclosed in the curve's loop. [*Hint: Use symmetry and horizontal rectangles*.]

c) Set up, but do not evaluate, an integral for the loop's length.

40. It's easy to find a parametric curve's **slope** at a given point. As always, Leibniz's notation does most of the work for us. We need only rewrite the usual expression for slope (dy/dx) as follows:

$$\frac{dy}{dx} = \frac{\frac{dy}{dt}}{\frac{dx}{dt}}.$$

Evaluating the right-hand side at the appropriate value of t immediately yields the slope.
With this in mind, answer the following questions about the curve shown in the previous problem.

a) When $t = 1/2$, where is the moving point? What is the curve's slope there?

b) Find the point or points on the curve at which the slope is exactly 1.

c) The moving point occupies the same position $(0,12)$ at two distinct times $(t = \pm\sqrt{3})$. Our slope formula yields different slopes at these times, even though the point's position on the curve is the same in both instances. What does this signify?

d) Of all the points on the curve's loop, one has the greatest x-coordinate. Obviously, the curve's slope at this point is infinite. Using this idea, find the point's coordinates.

e) If we revolve the loop around the y-axis, we obtain a tear-shaped solid. Find its volume.
[*Hint: Sketch a picture, find an expression for the volume of a typical slice, convert everything into terms of* t.]

41. Given a cycloid generated by a circle of radius a, find the slope of at the point where $\theta = 2\pi/3$.

42. Suppose one arch of the cycloid generated by a circle of radius a is revolved around the x-axis. Find the volume of the resulting solid.

Selected Answers
To Exercises

Chapter 1

1. The tangent is the line itself. **2.** The tangent will be perpendicular to the diameter containing P.

3. a) Measure the angle between the curves' *tangent lines* at the intersection point. b) 0

6. a) 0 b) $f'(2) < f'(6) < f'(3) < f'(4) < f'(1)$. c) 2 d) 4 e) $2, 6, 8$ f, g) TRUE

7. b, d, g are FALSE **8.** $h'(5) < g'(5) < f'(5)$ **9.** $f'(0) = 0,\ f'(5) = 0,\ f'(x) = 0$.

10. $g'(x) = a$ **11.** The domain of h' is all reals except 0. $h'(x) = \begin{cases} 1, & \text{if } x > 0 \\ -1, & \text{if } x < 0 \end{cases}$.

12. b) 0 c) $-1/\sqrt{3}$ d) -1 e) $-\sqrt{3}$

13. The graph of $h(t)$ is a parabola. At any time t, the book's speed corresponds to the magnitude of the parabola's slope at $(t, h(t))$. The parabola's symmetry ensures that at the two moments in question, these slopes must have equal magnitudes (but opposite signs). Hence, the two speeds will be equal.

14. a) The ball's vertical velocity 2 sec after leaving his hand. b) The time when the ball reaches maximum height. c) $p''(t)$ represents vertical acceleration; because the ball's acceleration is due only to gravity, it must be the case that $p''(t) = -32\ ft/s^2$. d) The ball is falling back to earth.

15. a) his velocity at time t. b) his speed at time t. c) his distance from the origin at time t. d) Yes, if he is east of the cigar, but traveling west. e) His velocity is *constant* between $t = b$ and c. f) He isn't moving between $t = d$ and e. g) At $t = n$, he changed his direction from west to east.

16. a) m/s^3 b) It is zero. **17.** The higher you go, the lower the temperature.

18. a) 160 minutes b) He will be moving faster at the beginning of the lap.

19. a) The rate at which the volume is changing after 5 hours, measured in ft^3/hour. b) $v'(t) < 0$ between $t = 24$ and 48.

20. $\frac{dy}{dx} = 3$ when $x = 2$; $\frac{dy}{dx} = e$ when $x = \pi$. **21.** $f'(-2) = 1,\ \ f'(\sqrt{5}) = \ln 5$

22. b) $f(5.002) \approx 8.0044$ c) $g(0.003) \approx 1.9955$ d) $\sin(3.19) \approx -0.05$ e) FALSE

23. a) $86\,J$ b) $\frac{dA}{dt} = -2\ m^2/min.$ when $t = 5$ min. c) $\frac{dz}{dt} = 4$ when $t = 6$. d) \$ / widgets per year.

24. b) dV/dh is greater when h is large. c) dV/dh would be greater when h is small. d) in increasing order: dV/dh when $h = 1,\ 17,\ 5,\ 10$. e) Any cylinder will do.

25. a) $\frac{d(\ln x)}{dx} = 5$ when $x = \frac{1}{5}$ b) $\frac{d(3x^3+1)}{dx} = 36$ when $x = 2$. c) $\frac{d(2^x)}{dx} = \ln 16$ when $x = 2$. d) $\frac{d(-4x+2)}{dx} = -4$. e) y is a linear function, so it has a constant derivative.

26. a) $x^2 - 2xdx$ b) $2xdx$ c) $x^3 + 3x^2dx$ d) $uv + udv + vdu$

27. a) $d(x^3) = (x + dx)^3 - x^3 = 3x^2dx$ [by exercise 26c]

28. a) $d\big(f(x)\big) = f(x + dx) - f(x)$. b) $d\big(5f(x)\big) = 5f(x + dx) - 5f(x)$.
c) $d\big(f(x) + g(x)\big) = \big(f(x + dx) + g(x + dx)\big) - \big(f(x) + g(x)\big)$.
d) $d\big(f(x)g(x)\big) = \big(f(x + dx)g(x + dx)\big) - \big(f(x)g(x)\big)$.
e) $d\Big(f\big(g(x)\big)\Big) = f\big(g(x + dx)\big) - f\big(g(x)\big)$.

29. a) $(3 + dx)^2 - 9 = \cdots = 6dx$. b) $\frac{d(x^2)}{dx} = 6$ when $x = 3$. c) $\frac{d(x^2)}{dx} = -1$ when $x = -\frac{1}{2}$.

31. 8 **32.** $y = -6x - 9$ **33.** a) $(-5/2,\ 25/4)$ b) $y = -5x - (25/4)$ **34.** $(2/3,\ 4/9)$

35. $2ax + b$ **36.** a) $3x^2$ b) $6,\ 3\pi^2$ c) $y = 3x - 2$ d) Yes, at $(-2, -8)$. **39.** $(nc)x^{n-1}$

40. a) $15x^2$ b) $-21x^6$ c) πx **41.** a) $9x^2 + 4x$ b) $-50x^4 + (3/4)x^2$ c) $\sqrt{2}$

43. a) $15x^2 - 8x^3$ b) $6x$ c) $-4x^9$ **44.** a) $6x^2 + 8x + 5$ b) $\frac{9}{4}x^2 - 18x - \sqrt{5}$ c) $-6x^5 + 3x^2 - 8x$

46. a) False. The power rule doesn't apply here, since the function $(2x + 1)^3$ isn't a pure power function; it is the composition of a linear function $(2x + 1)$ and a power function (x^3). b) $24x^2 + 24x + 6$

48. $\frac{dy}{dx} = 0$. **49.** Nothing at all, for 3^x isn't a power function; it is an *exponential* function.

50. a) $f'(t) = 6t^2 - 6t$ b) $g'(z) = z^3 + z^2 + z + 1$ c) $A'(r) = 2\pi r$ **52.** $y = 2x + 1$ **53.** $(-1, -1)$

54. The point moves *to the right* until $t = 2/3$, when it stops instantaneously before changing direction.

55. a) Point 1. b) when $t = 11/6$. c) When $t = 11/3$, both will be at position $-77/9$. Point 1's velocity will then be $4/3$ units/sec; Point $2's$ will be $-29/3$ units/sec. They will never meet again.

56. $79\ m/s$, $9.8\ m/s^2$ downwards.

60. a) dy/dx is a function, not an operator. Presumably, Dan's intended statement was $\frac{d}{dx}(x^3) = x^2$.

b) d/dx is an operator, not a function. He should have written: If $y = 2x^4$, then $dy/dx = 8x^3$.

c) Dan seems to be taking a derivative with respect to r, but he used the d/dx operator. Changing that x to an r would make his statement correct.

d) Here he has used dy where he presumably means dy/dx.

e) Equals abuse: The equals sign doesn't mean "has its derivative equal to". (It means "equals".)

f) An arrow should never be used where an equals sign is appropriate.

g) The arrow should be an equals sign. h) Dan's d should be $\frac{d}{dx}$. i) The $\frac{d}{dx}$ never goes on the right.

Chapter 2

1. a) $f'(\theta) = \cos\theta$ b) $\frac{dz}{dt} = -\sin t$ c) $\frac{dx}{dy} = \cos y$ d) $\frac{dB}{d\alpha} = -\sin\alpha$ e) $g'(w) = \cos w$

2. a) $f'(x) = 3\cos x + 5\sin x$ b) $y' = 2\pi t + \sqrt{2}\cos t$ c) $g'(x) = \sin x - \cos x$ **3.** 1

4. $(\pi/180)\cos\theta$ **7.** a) $y' = 3\cos(3x + 6)$ b) $y' = -6x\sin(3x^2)$ c) $y' = -\sin x\,[\cos(\cos x)]$

d) $y' = 150(3x + 1)^{49}$ e) $y' = 2\sin x\cos x = \sin(2x)$ f) $y' = 2\cos(2x)$

11. a) $y' = -24x^7\cos x + 3x^8\sin x$ b) $y' = \cos^2 x - \sin^2 x = \cos(2x)$

c) $y' = (36x^3 - 3x^2)\sin x + (9x^4 - x^3 + \pi^2)\cos x$

d) $y' = 2\pi x(\sin x + \cos x) + \pi x^2(\cos x - \sin x)$ **14.** $f'(x)g(x) + f(x)g'(x)$ **16.** $\frac{g(x)f'(x) - f(x)g'(x)}{[g(x)]^2}$

19. a) $y' = \frac{2}{(1-x)^2}$ b) $y' = \frac{9x^2\cos x + 3x^3\sin x}{\cos^2 x}$ c) $y' = \frac{\tan x - (x + \pi^2)\sec^2 x}{\tan^2 x}$ d) $y' = \frac{3x^2 + 10x}{(3x + 5)^2}$

e) $y' = \frac{4x + 2\sin(2x) + 8x^2\cos x}{(2x + \cos x)^2}$ f) $y' = \frac{1}{2}\sin(2x) + x\cos(2x)$ (double angle identities!)

g) $y' = \frac{x\sec x\tan x + x\sec^2 x - \sec x - \tan x}{x^2}$ h) $y' = -2x^{-3} + 3x^{-4} - 4x^{-5} + 5x^{-6}$

22. a) $y' = \frac{\sin x}{2\sqrt{x}} + \sqrt{x}\cos x$ b) $y' = \frac{2}{\sqrt{x}}$ c) $y' = \frac{\tan x - 2x\sec^2 x}{2\sqrt{x}\tan^2 x}$ d) $y' = 5\sqrt[3]{x^2}$

e) $y' = \frac{-\csc(\sqrt{x})\cot(\sqrt{x})}{2\sqrt{x}}$ **23.** $\left(\frac{6 - \sqrt{3}\pi}{6(2-\sqrt{3})}, \frac{6 - \sqrt{3}\pi}{6(2-\sqrt{3})}\right)$

24. a) $y' = e^x(\sin x + \cos x)$ b) $y' = (1/3x)(\cot x - x\csc^2 x\ln x)$

c) $y' = 2^x(\sin x - (\ln 2)\cos x) + 3x(1 + 2\ln x)$ d) $f'(x) = e^x\left(\frac{2x + 1}{2\sqrt{x}}\right)$

e) $g'(x) = (5\ln 120)120^x$ f) $y = 1 - 1 = 0$, so $y' = 0$.

g) $w' = \frac{5}{3}t^{2/3}\sec t + t^{5/3}\sec t\tan t - 1/t^2$ h) $\frac{dV}{dy} = \frac{y-1}{y^2}$ i) $k'(x) = 2^{x-1}[(\ln 4)x - 1]x^{-3/2}$

25. Yes, when $x = -\frac{\ln(\ln 2)}{\ln 2}$. **34.** $(f(g(x)))' = f'(g(x))g'(x)$

35. a) $y' = 5e^{5x}$ b) $y' = -6x\sin(3x^2)$ c) $y' = 1/(x\ln x)$ d) $y' = 8(2x^3 - x)^7(6x^2 - 1)$

e) $y' = (\ln 2)2^{\tan x}\sec^2 x$ f) $y' = -6(3x - 4)^{-3}$ g) $y' = \frac{144 - 2x^2}{\sqrt{144 - x^2}}$ h) $y' = -2xe^{-x^2}$

i) $y' = \frac{1 + 3x^2}{3(x + x^3)^{2/3}}$ j) $y' = -\sin(2x)$ k) $y' = 24x^2(1 - 4x^3)^{-3}$

l) $y' = -x^{-2}\sin(\csc(1/x))\csc(1/x)\cot(1/x)$ m) $y' = e^{\sin^2 x}\sin(2x)$ n) $y' = \frac{1}{x}\cot(\ln x)$

o) $y' = e^{\sin x}\left(\frac{2\cos x - \tan x}{2\sqrt{\sec x}}\right)$ p) $y' = \frac{2x^2+6x-8}{(x^2+4)(2x+3)}$ q) $y' = (\ln 10)10^{-x/(1+x^2)}\left[\frac{x^2-1}{(1+x^2)^2}\right]$

r) $y' = 10xe^{6x^2+e}(1+6x^2)$ s) $y' = \frac{4x+1}{6x^2+3x}$ t) $y' = 0$

36. $\left(-\frac{1}{9}, 0\right)$ **37.** There are infinitely many such points. One example is $(e^{\pi/2}, 0)$.

38. a) The same places $y = \sin x$ does. Namely, whenever x is an integer multiple of π.

b) The range is $[\sin -1, \sin 1]$, which is approximately $[-0.8514, 0.8415]$.

c) At the origin (and at all even integer multiples of π), the slope is 1; at odd multiples of π, it is -1.

d) The slopes at x-intercepts remain the same, but the range is compressed to $[-0.7456,\ 0.7456]$.

Chapter 3

1. Only statements *a-e* are true. **2.** a) $y = x^2$ b) $y = \sqrt{x}$ c) $y = (x-1)^2$ d) $y = 1 - x^2$.

4. b) The tangent to $y = x^3$ at the origin is horizontal, but there is neither hill nor valley there.

5. a) The second derivative is zero at the origin, which is not an inflection point. b) change sign there.

6. a) $-\sin x$ b) $2\sec^2 x \tan x$ c) e^x d) $(\ln \pi)^2 \pi^w$ e) $\cos t$ f) 2 g) True

7. If f is an nth-degree polynomial, then f' is a polynomial of degree $(n-1)$. As such, f' has at most $(n-1)$ zeros. Hence, f itself has at most $(n-1)$ turning points (hills or valleys) in its graph.

9. a) A "one-bump" cubic, could such a thing exist, would have just one turning point. Its graph, when viewed from afar, would either open upwards with a global minimum somewhere between its two "arms" that stretch up to infinity, or it would open downwards with a global maximum somewhere between its downward-stretching arms. Neither scenario is possible, since the range of any cubic is the set of *all* real numbers.

b) A cubic is a two-bumper if and only if $b^2 - 3ac > 0$. c) $-b/3a$ d) $(-3,0)$, $y = x^3 - 2x$

f) Applied to Bombelli's cubic, del Ferro's formula yields $x = \sqrt[3]{2 + \sqrt{-121}} + \sqrt[3]{2 - \sqrt{-121}}$.

Since $121 = 11^2$, we can also write this as $x = \sqrt[3]{2 + 11\sqrt{-1}} + \sqrt[3]{2 - 11\sqrt{-1}}$.

11. The field should be $125'$ by $250'$. **12.** Hint: Call the fixed perimeter P. **13.** $10 = \frac{\sqrt{61}-1}{3} + \frac{31-\sqrt{61}}{3}$

14. The radius should be $\sqrt[3]{1500/\pi} \approx 7.82$ cm. The height should be exactly twice this.

15. To *minimize* area, he should cut the wire so that its longer piece, which he'll bend into a triangle, will be $18/(9+4\sqrt{3}) \approx 1.130$ feet long. (Hence the short piece will be approximately 0.870 feet.) Assuming he must cut the wire, the best he can do to *maximize* the area is to cut the smallest piece possible that he can still bend into a triangle, and bend the rest into a square.

16. Nancy should make the angle $\pi/3$. **17.** Exactly $8/27$ (approximately 29.6%) of the sphere's volume

18. $\sqrt{5}$ units. **19.** a) \$85 b) \$110

21. a) $\frac{1-x^2}{y^2}$ b) $\frac{1+4x^3}{1-5y^4}$ c) $\frac{\sin x + x\cos x}{-\tan^2 y}$ d) $\frac{5x^4-3y}{3x+3y^2}$ e) $\frac{e^{x+5y}}{2y-5e^{x+5y}}$ f) $\frac{(\cos y)(\cos(x\cos y))}{1+x(\sin y)(\cos(x\cos y))}$

g) $\frac{1+y^2}{3x+3y-y^2-1}$ h) $\frac{y}{2\sqrt{xy}-x}$ **22.** Yes, at $(-1, -\sqrt[3]{2})$ and at $(1, \sqrt[3]{2})$.

23. It crosses at $(0,1)$ with slope $-1/4$, at $(0,0)$ with slope 1, and at $(0,-1)$ with slope $-1/4$.

24. It crosses at $(0,0)$ with slope 1, and at $(-1,0)$ with slope -3. **25.** The slope there is -1.

26. a) $\frac{dy}{dx} = \frac{x+1}{2-y}$ b) Hint: Rewrite the equation by completing the squares (in x and in y).

27. a) They cross at four points: $(\sqrt{3},\sqrt{3}), (-\sqrt{3},\sqrt{3}), (-\sqrt{3},-\sqrt{3}), (\sqrt{3},-\sqrt{3})$.

b) Find the two curves' slopes at an intersection point; verify they are negative reciprocals.

28. $5/4\pi$ cubits/min **29.** At that instant, the angle is decreasing at a rate of $2\sqrt{3}/15$ radians/sec.

30. 12 in²/sec. **31.** a) $\frac{dV}{dt}P + V\frac{dP}{dt} = 0$ b) 2.5 psi/min c) 62.5 psi/min **32.** 16π m²/sec.
33. b) 80π miles/min **34.** a) 7 ft/sec b) 3 ft/sec

Chapter π

2. Suppose $y = f(x)$ is a function. When we increase its input from x to $(x + \Delta x)$, its output changes from $f(x)$ to $f(x + \Delta x)$. Hence, it follows that

$$\frac{dy}{dx} = \lim_{\Delta x \to 0} \frac{\Delta y}{\Delta x} = \lim_{\Delta x \to 0} \frac{f(x + \Delta x) - f(x)}{\Delta x}.$$

5. a) 5 b) -8 c) does not exist d) ∞ e) 0 f) $\pi/2$ g) $-1/4$ h) $-6/\pi$
6. a) ∞ b) $-\infty$ c) ∞ d) 0 e) $-\infty$ f) $-\infty$ g) $-\infty$ **7.** 2 **8.** 2/3
9. a) 1 b) 3 c) 1/2 d) 1/2 e) 9/4 f) 2 **10.** a) 0 b) ∞ c) -3 d) 1/2 e) $-9/14$ f) 0
11. a) 1000 b) $\ln x = 1{,}000{,}000 \Rightarrow x = e^{1{,}000{,}000} > 2^{1{,}000{,}000} = (2^{10})^{100{,}000} > (10^3)^{100{,}000} = 10^{300{,}000}$, which has 300,000 digits. d) 0 e) The limit is 0, so $\sqrt{x}$ goes to infinity faster than $\ln x$.
13. a) 0 b) 0 c) ∞ d) 0 e) 1/3 f) 0

Chapter 4

2. a) $\int_0^3 f(x)dx$ b) $\int_{-\pi/3}^{\pi} \cos(x/2)\,dx$ c) $\int_0^2 (-y^4 + y^3 + 2y^2)dy$
3. a) No b) Yes. In any given instant, the particle's speed is constant, so $v(t)dt$ is the distance traveled during that instant. Hence, the *total* distance travelled is $\int_4^8 v(t)dt$.
4. b) A typical slice has volume $\pi x dx$. c) $\int_0^3 \pi x dx$ d) A typical slice has volume $\pi \sin^2 x\, dx$, so the solid's total volume is $\int_0^\pi \pi \sin^2 x\, dx$. e) A typical *horizontal* slice at y has radius $\sin^{-1} y$, and thus volume $\pi(\sin^{-1} y)^2 dy$. Hence, the solid's total volume is $\int_0^1 \pi(\sin^{-1} y)^2 dy$.
5. a) A typical slice's area is $(\sqrt{x} - x^2)dx$, so the total area is $\int_0^1 (\sqrt{x} - x^2)dx$. b) $\int_a^b (f(x) - g(x))dx$.
7. a) -5 b) 2 c) -3 d) 0 e) -5 f) 5 g) 11
8. a) The particle's displacement between $t = 0$ and $t = 2$. (In this particular case, since the velocity is always positive, an equivalent answer is: the total distance the particle travels during this time.)
b) The particle's displacement between $t = 0$ and $t = 5$.
c) The total distance the particle travels between $t = 0$ and $t = 5$. d) True e) False.
9. a) ½ b) 3 c) π/2 d) −π e) 3
10. a) All three have the form $(f(x) - g(x))dx$. b) $\int_a^c (f(x) - g(x))dx$. c) $\int_0^3 (3x - x^2)dx$
11. a) Some even functions: x^2, x^4, $|x|$, $\cos x$, $\sec x$. Some odds: x, x^3, $\sqrt[3]{x}$, $\sin x$, $\tan x$.
b) Evens are symmetric about the y-axis, odds are symmetric about the origin.
c,d) Think about the graphs. e) Odd. f) 0
12. b) $x^4/4$, $x^{10}/5$, $x^2/2$, $(-10/3)x^{3/2}$, $(3/5)x^{5/3}$, $-1/x$, $-14x^{4/7}$, $x^{1-\pi}/(1-\pi)$.
c) -1, since this would entail division by zero.
13. a) $\sin x$ b) $\cos x$ c) $-\cos x$ d) $\tan x$ e) e^x f) $\sec x$ g) $\arctan x$ h) $\cot x$ i) $\ln x$
14. a) $(1/5)\sin(5x)$ b) $(-1/\pi)\cos(\pi x)$ c) $-e^{-x}$ d) $e^{3x}/3$ e) $2\tan(x/2)$ f) $(1/\ln 10)10^x$
g) $(-1/\ln 5)5^{-x}$ h) $\frac{1}{2}\arctan(2x)$
15. a) They have distinct domains. b) $\ln|x|$. (Recall that this function's output will be $\ln x$ if its input x is positive, and $\ln(-x)$ if its input x is negative. Its domain is thus all reals except zero.)

16. a) 8 b) 0 c) 18 d) 0 **17.** 0 **18.** $A_{g,0}(x) = \int_0^x \cos t\, dt$. **19.** 8/9 units²

20. π/3 units² **21.** b) $\sin x$, $\sin(x^2)$, $2x\sin(x^2)$

22. a) Sentence 1 in paragraph 2. b) $dA_{f,a} = -(-f(x)dx)$, from which it follows that $dA_{f,a}/dx = f(x)$.

c) By assumption, $\boldsymbol{f(x) = 0}$. This also implies that the function's value is zero *in an infinitesimal neighborhood* of x, so by nudging x forward infinitesimally, we accumulate nothing new. That is, $dA_{f,a} = 0$ at x. From this, it immediately follows that $\boldsymbol{dA_{f,a}/dx = 0}$ at x. Equating the two boldface expressions equal to zero, we have $dA_{f,a}/dx = f(x)$ as claimed.

23. a) -1 b) -15/2 c) 2 d) π/12 e) $(\sqrt{3}-\sqrt{2})/2$ f) 45/4 g) $\sqrt{3}/6$ h) π/9 i) -2/5 j) $(2-\sqrt{2})/4$

24. a) 8/9 b) 3 c) 44/15 **25.** 96π/5 **26.** 128π/7 **29.** $\int_a^d f(x)dx$ **30.** a) -2 b) 3 c) 6

31. b) -1/5, $(1-e^{-5})/5$, 78, π/16 **33.** No. Counterexample: $\frac{1}{3} = \int_0^1 x^2 dx \neq \left(\int_0^1 x dx\right)\left(\int_0^1 x dx\right) = \frac{1}{4}$.

34. "continuous over $[a,b]$." A function is *continuous* over an interval if its graph can be drawn over that interval without lifting the pen from the page. This obviously isn't the case for the graph of $y = 1/x$ over $[-1,1]$ on account of its discontinuity at 0.

Chapter 5

1. a) $x^5 - 2x^4 + C$ [Hereafter, I'll omit the C to save space, but you should not.]

b) $\arcsin x$ c) $\frac{1}{5}\sin(5x)$ d) $\ln x$ e) $\ln(5+x)$ f) $5\arctan z$ g) $\frac{x^2}{2} + 2x + \frac{1}{x}$ h) $\frac{2}{5}x^{5/2} + \frac{6}{5}x^{5/6}$

i) $-\frac{1}{2}e^{-2t}$ j) $-\frac{1}{2}\ln(3-2x)$

2. a) The dx is missing. b) The C shouldn't be there. c) A definite integral is a number, not a function.

3. a) No. The function has infinitely many antiderivatives. b) Yes. In fact, $f(x) = x^3 - \frac{3}{2}x^2 + x + \frac{3}{2}$.

c) We initially knew that f had the form $f(x) = x^3 - (3/2)x^2 + x + C$ for some constant C. The graphs of the cubics in this family are vertically shifted versions of one another. Collectively, they cover the whole plane: Every point in the plane lies on one – and only one – of these cubics. Thus, specifying that the graph of f must pass through some specific point (such as (1,2)) reduces the infinitely many possibilities for f down to a single cubic, thus determining it completely. [You may (or may not) find it worthwhile to play around with a graphing program to get a better feel for it.]

4. a) $f(x) = \frac{2}{3}x^{3/2} + x^{5/2} - \frac{16}{3}$ b) $f(x) = 4x - \sin x + \cos x - 6\pi$ c) $f(x) = x^3 + 2x^2 - (1/2)x - (3/2)$

5. a) $h(t) = -16t^2 + 128t + 320$ b) 576 feet c) 10 seconds d) 192 ft/sec. **6.** 156′ 3″

7. $x(t) = -(1/4)\sin(2t) + 4t + 5$. **8.** 193.6′

9. Suppose F and G are antiderivatives of f and g respectively. Then you should be able to justify each equals sign in the following arguments:

a) $\int(f(x)+g(x))dx = \int(F'(x)+G'(x))dx = \int(F(x)+G(x))'dx = F(x)+G(x) = \int f(x)\,dx + \int g(x)dx$.

b) $\int cf(x)dx = \int cF'(x)dx = \int(cF(x))'dx = cF(x) = c\int f(x)dx$.

10. a) $\frac{1}{2}\sin^2 x$ b) $\frac{1}{6}(2x^2-1)^{3/2}$ c) $-2/(6x^2+1)$ d) 0 e) 1/3 f) $\ln(3x^3+2x+1)$ g) $\frac{1}{2}(\ln x)^2$

h) $\frac{1}{2}(\arcsin x)^2$ i) $2\sqrt{1+\sin x}$ j) $(1-e^{-4})/2$ k) $\arctan(e^x)$ l) -1/5 m) 2 n) $(\tan 1)^3$ o) $-2\cos\sqrt{x}$

11. According to the Antiderivative Lemma, any two antiderivatives of $\sin x \cos x$ must differ by a constant. Do these? Yes. They may not look like it at first, but playing around with the Pythagorean identity should convince you that David's antiderivative is Bathsheba's plus (1/2).

12. $-\ln(\cos x)$ **13.** $\ln(\sin x)$ **14.** a) $(\pi^2, 0)$ b) 4π c) 1

15. Letting $u = \sin x$ reduces the last integral to $\int[1/(1-u^2)]du$. **16.** Yes. $\cosh 0 = 1$.

17. $\sinh 0 = 0$. $\frac{d}{dx}(\sinh x) = \cosh x$. **18.** Yes. $A_{x^2,2}(x) = \frac{1}{3}(x^3 - 8)$

21. a) $3\ln|x+5|$ b) $\frac{2}{5}\ln|5x-3|$ c) $3/2$

22. a) $\frac{3}{x-1} + \frac{2}{x+3}$ b) $\frac{3}{x-3} - \frac{2}{x-2}$ c) $\frac{1}{x} + \frac{3}{x-2} - \frac{1}{x+2}$ **23.** a) $1 + \frac{5}{x-1}$ b) $2 + \frac{2x-4}{3x^2-x+1}$ c) $4x^2 - 5 + \frac{16}{2x^2+3}$

25. a) $3\ln|x-1| + 2\ln|x+3|$ b) $2\ln|x| - 4\ln|x+5|$ c) $\frac{x^3}{3} + 4x + 4\ln\left|\frac{2-x}{2+x}\right|$ **26.** $\ln\left|\frac{P}{1-mP}\right|$

28. $-\ln|\cot x + \csc x|$ **29.** Divide both sides of the main Pythagorean identity by $\cos^2\theta$.

30. You didn't read the section, did you? Break this bad habit.

31. b) $(\pi\cos^2 x - \pi\sin^2 x)dx$ [or better yet, $\pi\cos(2x)\,dx$.] c) $\pi\int_0^{\pi/4}\cos(2x)\,dx$ d) $\pi/2$ units³.

32. a) $\frac{1}{2}(x + \sin x\cos x)$ b) $\frac{1}{4}(\pi + 2\sqrt{2} - 4)$ c) $\frac{1}{8}[4\theta + \sin(4\theta)]$ d) $\frac{1}{96}(3\sqrt{3} + 8\pi)$ e) $\frac{1}{2}\sin(2x)$ f) $\sqrt{3} - 1$

g) 0 h) $\frac{\cos^5 x}{5} - \frac{\cos^3 x}{3}$ i) $\frac{\cos^9 x}{9} - \frac{\cos^7 x}{7}$ j) $\frac{\sin^4 x}{4} - \frac{\sin^6 x}{6}$ k) $\frac{1}{32}[4x - \sin(4x)]$ l) $\frac{\tan^8 x}{8} + \frac{\tan^6 x}{3} + \frac{\tan^4 x}{4}$

33. b) $\left(\frac{\pi}{4}, 1\right)$, $\left(\frac{\pi}{3}, \sqrt{3}\right)$, $\left(\frac{\pi}{6}, \sqrt{3}\right)$ d) $\pi(3 - \cot^2 x)dx,\ \pi(3 - \tan^2 x)dx$ e) $2\pi\int_{\pi/4}^{\pi/3}(3 - \tan^2 x)dx$

f) $\frac{2\pi}{3}(\pi + 3 - 3\sqrt{3}) \approx 1.98$ units³.

34. a) Letting $x = 3\sin\theta$ (or $3\cos\theta$) turns the orginial expression into $9\cos^2\theta$ (or $9\sin^2\theta$).

b) Letting $x = \sqrt{2}\sin\theta$ (or $\sqrt{2}\cos\theta$) yields $2\cos^2\theta$ (or $2\sin^2\theta$).

c) Letting $x = 4\tan\theta$ yields $16\sec^2\theta$. d) Letting $x = \sqrt{15}\tan\theta$ yields $15\sec^2\theta$.

e) Letting $x = \sqrt{3}\sec\theta$ yields $3\tan^2\theta$. f) Letting $x = \frac{5}{3}\tan\theta$ yields $25\sec^2\theta$.

35. a) $\frac{-\sqrt{9-x^2}}{x} - \arcsin\left(\frac{x}{3}\right)$ b) $-\frac{2}{3}\sqrt{9-x^3}$

c) $\ln\left|\sqrt{4+x^2} + x\right| - \ln 2 + C$, or better yet, $\ln\left|\sqrt{4+x^2} + x\right| + C$, since the constant term $\ln 2$ can be absorbed into the constant of integration.

d) $\sqrt{x^2 - 4}$ e) $\arcsin(x/4)$

36. By a well-known property of logarithms, $\ln\left|\frac{x}{3}\right| + C = \ln|x| - \ln 3 + C = \ln|x| + (C - \ln 3)$.

Thus, our complete set of antiderivatives consists of all functions of the form $\ln|x| + (C - \ln 3)$, where C can be any real number. But as C ranges over all the reals, the expression $(C - \ln 3)$ takes on *all real values*. (For example, it will take on the value 5 when $C = (5 + \ln 3)$.) Consequently, the full set of antiderivatives could be described more simply as the set of all functions of the form $\ln|x|$ + (a constant).

37. $\ln\left|\sqrt{x^2 - 2} + x\right| + C$ **39.** $\frac{1}{4}\arctan\left(\frac{x}{4}\right) + C$

40. a) $e^x(x-1)$ b) $\frac{x^{10}}{100}[10\ln(x) - 1]$ c) $2x\sin x - (x^2 - 2)\cos x$ d) $\frac{1}{9}[\sin(3x+2) - 3x\cos(3x+2)]$

e) $\ln x\,[\ln(\ln x) - 1]$

41. a) $x\ln x - x$ b) $x\arctan x - \frac{1}{2}\ln(x^2+1)$ c) $x\arcsin x + \sqrt{1-x^2}$ d) $x\arccos x - \sqrt{1-x^2}$

42. The "sentence": $\int e^x\cos x\,dx = e^x\sin x - \int e^x\sin x\,dx = e^x\sin x - (-e^x\cos x + \int e^x\cos x\,dx)$.

From this, we extract this equation: $\int e^x\cos x\,dx = e^x\sin x + e^x\cos x - \int e^x\cos x\,dx$.

Solving for the integral, we reach our conclusion: $\int e^x\cos x\,dx = \frac{1}{2}e^x(\sin x + \cos x)$.

43. $2\sqrt{1-x^2}\arcsin x - 2x + x(\arcsin x)^2$ **44.** $2\pi^2$ units³. **45.** d) $\frac{48\pi}{5}$ units³.

Chapter 6

1. a) $14/3$ units b) $\ln(\sqrt{2} + 1)$ units **2.** d) $14/3$ units

3. a) $\frac{\pi}{27}(145\sqrt{145} - 1)$ units². b) $\frac{\pi}{6}(17\sqrt{17} - 1)$units².

5. a) $\int_0^{\pi} 2\pi \sin x \sqrt{1+\cos^2 x}\, dx$ b) $\int_0^1 2\pi e^{-x}\sqrt{1+e^{-2x}}dx$ **6.** 14.424

7. d) $\int \sec^3 x\, dx = \frac{1}{2}\sec x \tan x + \frac{1}{2}\ln|\sec x + \tan x| + C$ **8.** $\frac{1}{4}\left[2\sqrt{5} + \ln\left(2+\sqrt{5}\right)\right]$ units

11. a) $\sqrt{2}\pi$ units². b) A cylinder at x would have lateral surface area $2\pi x dx$; the sum of all their areas is π units².

12. a) The trapezoid consists of the rectangle topped by a triangle. But the triangle has zero area (its area is a second-order infinitesimal, $(dx \cdot dy/2)$), so it adds nothing to the area of the rectangle. Accordingly, when finding the area under the curve, we can just integrate the rectangles without having to worry about the trapezoids.

b) Conical frusta will work for both volumes and surface areas. This follows from the first basic principle of calculus (curves are locally straight), which tells us that the solid's generating curve is a chain of infinitely many infinitesimally small line segments. It follows that the area beneath the curve consists of infinitely many infinitesimally thin trapezoids. As the curve (and the area beneath it) revolves, the trapezoids trace out conical frusta. The solid is thus the infinite collection of these infinitesimally-thin frusta. Hence, its volume and surface area must equal the sums of the volumes and surface areas of its constituent parts: the conical frusta.

c) A solid frustum consists of a cylindrical core with a "collar" wrapped around it. The collar is itself a solid of revolution generated by an infinitesimal triangle. If we cut the collar and "unroll" it, it becomes a long triangular prism. This prism's triangular base has zero area (a second-order infinitesimal), so the prism's volume, which is the collar's volume, is also zero. Since the collar adds nothing to the volume of the frustum's uncollared cylindrical core, we can, when finding the volume of a solid of revolution, simply integrate the cylinders without having to worry about the frusta.

d) Something specific on which you might meditate: In part (a), the rectangle was literally a part of the trapezoid. In part (c), the solid cylinder was literally part of the solid frustum. In contrast to these cases, the lateral surface of the cylinder is *not* part of the frustum's lateral surface.

13. a) 1 c) $m = 3$; $2/3$ d) $n = 485{,}165{,}196$; 0.999999998 **14.** a) 1 b) Divergent c) $1 - \cos 1$ d) $\pi/6$

15. a) $\int_{-\infty}^{b} f(x)dx = \lim_{a\to-\infty} \int_a^b f(x)dx.$ b) π d) $y = \frac{1}{\sigma\sqrt{2\pi}} e^{\frac{-(x-\mu)^2}{2\sigma^2}}$

16. a) The volume is π units³. b) $2\pi \int_1^{\infty} \frac{1}{x}\sqrt{1+\frac{1}{x^4}}dx > 2\pi\int_1^{\infty}\frac{1}{x}dx$, which we've shown is infinite.

17. a) $\Gamma(1) = \Gamma(2) = 1$ c) $\Gamma(3) = 2,\ \Gamma(4) = 6,\ \Gamma(5) = 24,$ and in general, $\Gamma(n) = (n-1)!$

19. a) 34 b) 108 c) 0 **20.** a) $\sum_{i=1}^{100} i$ b) $\sum_{i=2}^{20} 1/i^2$ c) $\sum_{i=1}^{7} \frac{(-1)^{i+1}}{i!}$ **21.** $\sum_{i=1}^{n} i = \frac{n(n+1)}{2}$

22. a) $\square_n = n^2,\ \Delta_n = \frac{n(n+1)}{2}$ b) Yes. An example is 36. (Another is 1225.) d) 28 is perfect.

e) The next three Mersenne primes after 3 are 7, 31, and 127.

23. a) 0.285 b) 0.385 c) 0.3325

24. a) It first appears as a point, which then expands into a growing sphere; when its radius reaches 2 feet, it begins to contract again, shrinking down to a single point again, before it disappears. b) $8\pi^2$ feet⁴.

25. A sphere of radius $\sqrt{3}/2$ contained inside the hyperplane $x = 1/2$.

26. a) Look at the lines along which the walls, ceiling, and floor intersect. You'll find skew lines.

b) One way: Let l be the first randomly chosen line. Set up axes so that l lies in the xy-plane. Unless the second randomly chosen line, m, happens to be parallel to that plane (unlikely, given that m is chosen at random), it will pierce it at one - and only one - point. Unless that point happens to be on the x-axis (another unlikely event), lines l and m will never meet. Thus, for two randomly drawn lines in space, parallelism and intersection are freak occurrences – exceptions to the rule. Skew lines are the norm.

27. $2\pi^2 r^3$.

28. a) $x^2 + y^2 + z^2 + w^2 + u^2 = r^2$ b) 4D hypersphere of radius $\sqrt{r^2 - k^2}$. c) $(8/15)\pi^2 r^5$ d) $(8/3)\pi^2 r^4$

29. a) 78.5% b) 52.4% c) 30.8% d) 16.4% e) The percentage approaches zero.

Chapter 7

1. a) $x - \frac{x^3}{3!} + \frac{x^5}{5!} - \frac{x^7}{7!}$ b) $1 - \frac{x^2}{2!} + \frac{x^4}{4!} - \frac{x^6}{6!}$ c) $1 + x + \frac{x^2}{2!} + \frac{x^3}{3!} + \frac{x^4}{4!} + \frac{x^5}{5!} + \frac{x^6}{6!} + \frac{x^7}{7!}$
d) $1 + x + x^2 + x^3 + x^4 + x^5 + x^6 + x^7$

2. $x + \frac{x^3}{3} + \frac{2x^5}{15}$ **3.** The Taylor polynomial is $p(x)$ itself.

4. b) $1 - \frac{1}{2}\left(x - \frac{\pi}{2}\right)^2 + \frac{1}{24}\left(x - \frac{\pi}{2}\right)^4 - \frac{1}{720}\left(x - \frac{\pi}{2}\right)^6$ c) $2 + 2(x - \ln 2) + (x - \ln 2)^2 + \frac{1}{3}(x - \ln 2)^3$

5. a) If f is even, its graph is symmetric about the y-axis. In particular, for all x in f's domain, the tangent lines at x and $-x$ are mirror images of one another, reflected across the y-axis. Their slopes are thus equal in magnitude but opposite in sign. That is, $f'(x) = -f'(x)$ for all x in f's domain, so f' is odd.

b) If f is odd, its graph remains the same after successively reflecting it across *both* axes. In particular, the graph's tangent at x, after being reflected across both axes, will coincide with the graph's tangent at $-x$. Since each reflection changes the sign of the tangent's slope, the *two* reflections ensure that the tangent line's slope at $-x$ will be the same as it is at x. That is, $f'(-x) = f'(x)$, so f' is even.

c) Obviously, $f(-0) = f(0)$. By f's oddity, we also have $f(-0) = -f(0)$. Hence, $f(0) = -f(0)$, which can only be true if $f(0) = 0$. Hence, f's graph passes through the origin as claimed.

d) If f is even, then its odd-order derivatives must be odd functions (by parts a and b), and these necessarily vanish at zero (by part c), so f's Taylor polynomial can only contain even-power terms. A similar argument holds if f is odd.

6. Answers may vary. Hints for one path: Since $e < 3$, we know that $\sqrt{e} < \sqrt{3} < 2$. Using this bound as (part of) the expression that emerges from the Error Bound Theorem, you can deduce that the error must be less than $.00005$. From this, you can then conclude that $1.6486 < \sqrt{e} < 1.6488$.

7. Increasing the polynomial's degree makes a much bigger difference.

8. The coefficient for the x^6 term in sine's Taylor expansion is zero. Since $|E_6(0.3)| < .00000005$, we may conclude that sin (0.3) lies between 0.2955202 and 0.2955203.

9. a) 0 b) 0 c) ∞ d) 0

11. a) The coefficients for a Taylor polynomial for f centered at zero involve derivatives of f evaluated at zero, but at zero, the derivatives of this particular function are *undefined*.

b) The polynomial is $1 + (1/2)(x - 1)$. It gives the approximation $\sqrt{1.04} \approx 1.02$. Regarding the error bound, once you've graphed $f''(x)$ correctly (and correctly understood the statement of the EBT), you should see that an appropriate value of M is $1/4$.

c) The polynomial is $1 + x/2$.

13. a) $1 + x + \frac{x^2}{2!} + \frac{x^3}{3!} + \cdots$ b) $1 + x + x^2 + x^3 + \cdots$ c) $x - \frac{x^3}{3!} + \frac{x^5}{5!} - \frac{x^7}{7!} + \cdots$
d) $1 - \frac{x^2}{2!} + \frac{x^4}{4!} - \frac{x^6}{6!} + \cdots$ e) $x - \frac{x^2}{2} + \frac{x^3}{3} - \frac{x^4}{4} + \cdots$ f) $1 + \alpha x + \frac{\alpha(\alpha-1)}{2!}x^2 + \frac{\alpha(\alpha-1)(\alpha-2)}{3!}x^3 + \cdots$

14. (Note: Variations are possible, especially when the terms alternate between positive and negative.)
a) $\sum_{n=0}^{\infty} \frac{x^n}{n!}$ b) $\sum_{n=0}^{\infty} x^n$ c) $\sum_{n=0}^{\infty} \frac{(-1)^n}{(2n+1)!} x^{2n+1}$ d) $\sum_{n=0}^{\infty} \frac{(-1)^n}{(2n)!} x^{2n}$ e) $\sum_{n=1}^{\infty} \frac{(-1)^{n+1}}{n} x^n$

16. a) Geometric. Sum is 3. b) Not geometric. c) Geometric, but divergent. d) Not geometric.
e) Geometric. Sum is $\pi^2/(\pi - 1)$ f) Geometric, but divergent.

20. a) Sum is 6. b) Divergent. c) Sum is $-2/3$. d) Divergent.

21. Both arguments operate with divergent series as though they represent numbers. They don't.

22. a) $|x| < 1$ c) $1 + 3x + 9x^2 + 27x^3 + \cdots = 1/(1 - 3x)$, for all $|x| < 1/3$.
d) $1 - 4x^2 + 16x^4 - 64x^6 + \cdots = 1/(1 + 4x^2)$, for all $|x| < 1/2$.

24. a) Converges to 2. (Geometric series, $|r| < 1$) b) Diverges (Divergence test: general term goes to 2, not 0.) c) We can't tell yet, though we'll be able to answer the question soon. [The important thing to recognize here is that even though the general term vanishes, that *isn't* enough to guarantee convergence. Remember the harmonic series!] d) Diverges (Harmonic series) e) Diverges (This is still the tail of the harmonic series, and it is the infinite tail that determines convergence or divergence.) f) Diverges. (Geometric series, $|r| > 1$.) g) Diverges (Divergence test. General term goes to 1, not 0.) h) Diverges (Divergence test. General term goes to 2/9, not 0.) i) We can't tell yet, but that will soon change. j) Diverges. (Divergence test: General term doesn't converge; it oscillates between 2 and -2. Since the general term doesn't vanish, the series diverges.)

25. a) 31 terms b) 83 terms to pass 5; 227 terms to pass 6; 12367 terms to pass 10. c) first billion terms add up to about 21. The first trillion add up to about 28. The first 10^{82} terms add up to about 189. e) It would take about 317 years. Instead, your computer is most likely approximating the sum by using properties of the famous (in some circles) irrational number γ, which is linked to the harmonic series and the natural logarithm. Gamma (sometimes called "Euler's gamma") is perhaps the third most famous special irrational number, after π and e.

26. It's a valid proof, and a clever one at that.

28. Hint: Let the area of the full square be 1 unit^2. Then the areas of the successive L-shapes are... ?

29. a) We're just pulling out a common factor: $\sum ca_n = ca_1 + ca_2 + \cdots = c(a_1 + a_2 + \cdots) = c\sum a_n$.
b) Similar story. We're just reordering the terms.
[Note: The preceding justification will make pedants foam at the mouth and object that reordering an infinite series' terms sometimes isn't allowed. After reading about **absolute convergence** later in Chapter 7, you'll understand their objection. You'll then want to return to to this question and find a less intuitive – but perfectly rigorous – justification. (The key: The limit definition of a series.) But until you can understand the objection, the intuitive if flawed argument will serve.]

30. a) $\sum(-n)$ stays below the convergent p-series $\sum 1/n^2$, but it diverges (by the divergence test).
b) $\sum(-1/2)^n$ stays above the divergent series $\sum(-n)$, but it is a convergent geometric series.

31. a) convergent p-series b) divergent p-series c) divergent geometric series d) dvgt p-series
e) cvgt p-series f) cvgt geometric series g) cvgt by direct comparison with $\sum 1/n^2$
h) cvgt by limit comparison with $\sum 1/n^2$ i) dvgt since it is a multiple of the harmonic series
j) cvgt by direct comparison with $\sum 1/2^n$ k) dvgt by direct comparison with $\sum 1/\sqrt{n}$
l) cvgt by comparison with $\sum 1/n^{3/2}$ (That is, $\sum n^2/\sqrt{n^7+1} \le \sum n^2/\sqrt{n^7} = \sum 1/n^{3/2}$)
m) cvgt by limit comparison with $\sum 1/n^2$ n) dvgt by divergence test
o) cvgt by direct comparison with $\sum 1/2^n$ p) cvgt by direct comparison with $\sum 1/n^2$
q) cvgt by limit comparison with $\sum 1/n^4$ r) cvgt by limit comparison with $\sum 1/n^2$
s) dvgt; it is a multiple of the harmonic series t) dvgt by direct comparison with the harmonic series.

32. The inequality reverses (and remains reversed) at the fifth term, so it doesn't hold in the tails of the series. Consequently, the comparison test does not apply. In fact, a limit comparison test with the harmonic series shows that the left series *diverges*.

34. a) In the limit, a_n is tiny relative to b_n. Hence, in the tail, the terms of $\sum a_n$ stay below those of the convergent series $\sum b_n$. Thus, $\sum a_n$ converges by the direct comparison test.

35. a) converges by Leibniz' test. b) converges by Leibniz' test. c) Leibniz' test doesn't apply; this isn't an alternating series. (It's a convergent p-series.) d) Leibniz' test doesn't apply; the terms of this alternating series don't vanish. e) Leibniz' test doesn't apply; terms vanish, but not monotonically. f) For every value of x, this series – the Taylor series for cosine – converges by Leibniz' test.

38. a) abs cvgt b) abs cvgt c) abs cvgt d) cond cvgt e) dvgt f) cond cvgt (cf. exercise 31t)
g) abs cvgt h) dvgt i) dvgt

40. b) Suppose, by way of contradiction, that it converges. Then $\sum a_n = s$ for some number s. We know already (from part (a)) that $\sum b_n$ converges to some number t, so it would then follow that

$$s - t = \sum a_n - \sum b_n = \sum(a_n - b_n) = \sum 1/n.$$

That is, **if** $\sum a_n$ converges to s, then the harmonic series converges to $s - t$, which we know to be false. Hence, $\sum a_n$ must diverge, as claimed.

43. a) cvgt b) cvgt c) dvgt d) dvgt e) cvgt f) ratio test fails. (The series converges, though, by limit comparison with $\sum 1/n^2$.)

45. We're not interested in the ratio per se; we want its *limit* as $n \to \infty$. In either formulation (term-to-predecessor or successor-to-general-term) the ratio approaches the same limiting value. If you really wanted to be perverse, you could even use, say, $|a_{n+42}/a_{n+41}|$. All that matters is that we have a way to examine, in the limit, the ratio between successive terms.

46. a) If $\sqrt[n]{|a_n|} \to r$, then $|a_n| \to r^n$. That is, the nth term of the series $\sum |a_n|$ is, in the tail, effectively indistinguishable from r^n. Thus, in the tail, the series $\sum |a_n|$ looks like $r^n + r^{n+1} + r^{n+2} + \cdots$, which is geometric, and thus converges when $r < 1$ and diverges when $r > 1$.

b) i. Divergent, ii. Convergent

47. b) both series converge by the integral test. (Use integration by parts to get the second one.)

48. a) $|x| = \begin{cases} x, & if\ x > 0 \\ -x, & if\ x < 0 \end{cases}$ b) The distance of x from the origin on the number line.

c) The statement is obvious if $a > b$ or if $a = b$. If, however, $a < b$, then $a - b < 0$, so by part (a), $|a - b| = -(a - b) = b - a$, which is clearly the distance between a and b in this case. Thus in all possible cases, $|a - b|$ is the distance between a and b, as claimed.

d) By part (b), this is the set of all points less than a units from the origin. i.e. the interval $(-a, a)$.

e) By part (c), this is the set of all points whose distance from a is less than b. Draw a picture, and it will be clear that this set is the interval $(a - b, a + b)$.

f) Think about part (d).

g) Apply part (f) to the initial inequality. Then add a across all parts. (Or just think geometrically.)

h) $(-2,2)$ or $-2 < x < 2$; $[7/3, 11/3]$ or $7/3 \le x \le 11/3$; $(-2,3)$ or $-2 < x < 3$

49. No.

50. a) $[-1,1]$ b) $[-1,1]$ c) $[-1,1)$ d) $(-\sqrt{5}, \sqrt{5})$ e) $(-1/3, 1/3]$ f) $(2,6)$ g) $(-\infty, \infty)$ h) $(-1,1)$ i) $(-\infty, \infty)$ j) $(0, 2e)$

51. b) No. Any power series $a_0 + a_1(x - a) + a_2(x - a)^2 + \cdots$ will certainly converge at its center, for when $x = a$, the series becomes $a_0 + a_1(a - a) + a_2(a - a)^2 + \cdots$, whose sum is clearly a_0.

52. a) A Taylor series is a power series whose coefficients have the form $f^{(n)}(a)/n!$, for some function f.

b) $(-\infty, \infty), (-\infty, \infty), (-1,1), (-1,1]$.

53. The problem occurs in the penultimate line. $\sum |a_n(c - a)^n|$ is NOT known to be convergent. We are assuming only that $\sum a_n(c - a)^n$ converges. If it happens to converge *conditionally*, then $\sum |a_n(c - a)^n|$ would diverge, and the comparison test that follows would, of course, be invalid.

54. a) $1 - x + x^2 - x^3 + \cdots$ (r.o.c: 1) b) $1 - 16x^4 + 256x^8 - 4096x^{12} + \cdots$ (r.o.c: ½)

c) $x^2 - (x^6/3!) + (x^{10}/5!) - (x^{14}/7!) + \cdots$ (r.o.c: ∞) d) $1 - x^2 + x^4/2 - x^6/6$ (r.o.c: ∞)

55. $\ln(1 + x) = x - x^2/2 + x^3/3 - x^4/4 + \cdots$ **56.** $\ln 2$

57. $\int_0^1 e^{-x^2} dx \approx \int_0^1 (1 - x^2 + x^4/2 - x^6/6) dx = 26/35$. [The error here is about 0.004.]

58. πi. (Actually, there's more to the story... but you'll need to study complex analysis for that.)

59. a) $(1/\sqrt{2}) + (1/\sqrt{2})i$ b) $(\pi/2)i$ **61.** a) 2 b) $\sqrt{5}$ **62.** a) 1 b) 2

63. c) $\arcsin x = x + \left(\frac{1}{2}\right)\frac{1}{3}x^3 + \left(\frac{1\cdot 3}{2\cdot 4}\right)\frac{1}{5}x^5 + \left(\frac{1\cdot 3\cdot 5}{2\cdot 4\cdot 6}\right)\frac{1}{7}x^7 + \cdots$

Chapter 8

1.

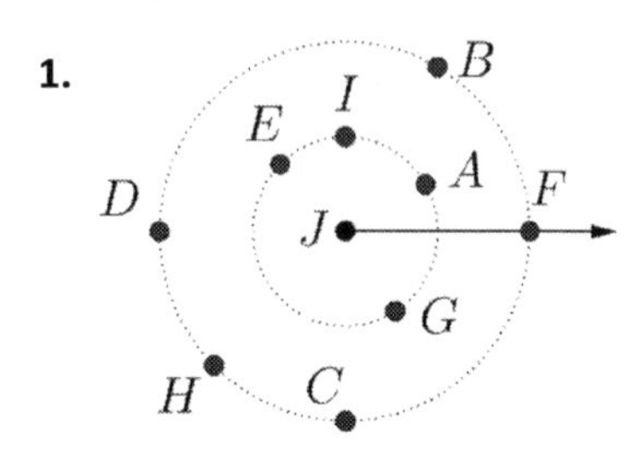

2. a) $(3,\pi/2)$ b) $(3,0)$ c) $(3,\pi)$ d) $(3,3\pi/2)$ e) $(\sqrt{2},\pi/4)$ f) $(2\sqrt{2},5\pi/4)$ g) $(2,\pi/3)$

3. a) A circle with radius 2 centered at the origin b) A ray from the origin, making a $45°$ angle with the polar axis. c) The line containing the ray in the previous part. d) The interior of the unit circle. e) The first quadrant.

4. a) $\theta = 0,\ r \geq 0$ b) $\theta = 0$ c) $r = 3$ d) $1 < r < 3$

5. Check your work with a graphing program. (As of this writing, Desmos is freely available online, and handles polar graphs with ease.)

6. Yes. Any point with coordinates $(-1,\theta)$ lies on the graph.

7. a) There's more. b) Hint: Consider the value of $\ln\theta$ when θ is tiny. c) No, this is not the case. There is some space between any two layers of the "wrap", though this space gets smaller with each successive layer. d) Hint: As θ goes to infinity, so does $\ln\theta$, but what about the *derivative* of $\ln\theta$? e) You are on your own.

8. a) $(-5,0)$ b) $(-5\sqrt{3}/2,-5/2)$ c) $(0,0)$ d) $(5\sqrt{2},-5\sqrt{2})$ e) $(1,-\sqrt{3})$

9. a) $(3,0)$ b) $(2,5\pi/3)$ c) $(\sqrt{2},3\pi/4)$ d) $(2,7\pi/6)$

10. Basic right-angle trig shows that $-x = r\cos(\pi-\theta)$. The RHS equals $-r\cos\theta$ by a basic identity, so $x = r\cos\theta$, as claimed. Similar story for $y = r\sin\theta$. The Pythagorean Theorem immediately gives $r^2 = x^2 + y^2$. Basic trig gives $\tan(\pi-\theta) = -y/x$. The LHS is $-\tan\theta$ by a simple identity, $\tan\theta = y/x$, as claimed.

11. The formula $\tan\theta = y/x$ fails for points on the vertical axis, since for such points, both sides of this conversion formula are undefined. (In practice, this does no harm; if a point lies on the vertical axis, we can convert its coordinates from one system to the other by just thinking about it. We don't need a formula.)

12. a) $r = 2$ b) $r = 2\csc\theta$ c) $r = 5/(2\cos\theta + 3\sin\theta)$ d) $r^2 = 9\cos 2\theta$ e) $r^2 = \tan\theta$

13. a) $x = -1$ b) $y = 3$ c) $x^2 + y^2 = 5x$ d) $x^2 + y^2 = 4$ e) $y^2 + 4x = 4$ f) $y = x/\sqrt{3}$ **14.** 1

15. An even polar function's graph is symmetric about the horizontal (polar) axis. An odd polar function's graph is symmetric about the vertical axis (i.e. the line $\theta = \pi/2$).

16. 6π **18.** $(5\pi - 6\sqrt{3})/24$ **19.** 2

20. a) $2 - \pi/2$ b) The arms approach the line $y = 2$ asymptotically. To see this, note that the y-coordinate of a general point on the graph is given by $y = \tan(\theta/2)\sin\theta$. Taking the limit as θ goes to π should answer our question, but the limit leaves us with an $\infty \cdot 0$ "tug of war". The way out of this is to rewrite the expression for y as $\sin(\theta)\,/\cot(\theta/2)$. In this form, the limit involves a $0/0$ tug of war, which you can handle with L'Hôpital's rule.

21. $\pi/2 - 1$

22. Equating the expressions for r and solving the resulting equation (in θ) leads us, in principle, to all pairs (r,θ) that satisfy *both* curves' equations. The problem here is that no single polar representation of the origin satisfies both equations; the origin lies on the circle because it is satisfied by $(0,\pi/2)$, while it lies on the heart because it is satisfied by $(0,\pi)$.

24. $\frac{3\sqrt{3}-\pi}{\pi\sqrt{3}+3}$ **25.** $\sqrt{3}/2$

26. a) A: $\left(\frac{1}{2},\frac{\pi}{3}\right)$, B: $\left(\frac{-\sqrt{3}}{2},\frac{-2\pi}{3}\right)$ b) A: $\left(\frac{-1}{2},\frac{-2\pi}{3}\right)$, B: $\left(\frac{\sqrt{3}}{2},\frac{\pi}{3}\right)$ c) Slope at A is $-5/\sqrt{3}$, at B it is $-\sqrt{3}/7$. e) $\pm 1/2$ f) $\pm 2\sqrt{6}$

29. a) The line $y = \frac{x}{2} + 1$ b) The parabola $x = y^2/4$ c) The portion of $y = 2x$ that lies in the first quadrant.

30. a) $x = t, \quad y = f(t)$. b) $x = t, \quad y = \sin(1/t)$. c) $x = t, \quad y = t \tan\theta$.

31. a) As t ranges from $-\infty$ to ∞, the point traces out the line $y = x$ from left to right and moves at a constant speed ($\sqrt{2}$ units per unit of time). At $t = 0$, the point is at the origin.

b) Same as part (a), but now the point's speed varies; the further from the origin, the faster it moves.

c) When $t = 0$ the point is at $(2{,}2)$. It moves left to right at a constant speed of $3\sqrt{2}$ units per unit of time.

d) When $t = 0$ the point is at $(1{,}1)$. It moves *right to left* at a constant speed of 2 units per unit of time.

e) The point emerges from the origin (when $t = 0$), heads right until it reaches $(\pi/2\,, \pi/2)$, turns around, passes back through the origin until it reaches $(-3\pi/2\,, -3\pi/2)$, then turns around again. The point continues this back-and-forth behavior, venturing a bit further from the origin each time.

32. b) $x = 2\cot\theta\,, \quad y = 2\sin^2\theta$

34. Hint: Recall the conversion formulas from polar to rectangular coordinates.

35. a) The line through $(2{,}4)$ with slope $-2/3$. b) The line in space passing through points $(2{,}4{,}5)$ and $(5{,}2{,}6)$.

c) A spiral in the plane. d) A "tornado" in space.

36. a) $2\pi a$ units b) At point O, the value of θ is 0; at point Q, the value of θ is 2π.

37. a) The values of t at the right, top, left, and bottom cusps are $0,\ \pi/2\,,\ \pi,\ 3\pi/2$ respectively.

b) $ds = 3\cos t \sin t$. c) The portion in the first quadrant has length $3/2$, so the total length is 6 units.

d) The integral in this part is equal to 0. As for the explanation of why this integral does not give the arclength as expected, here is a hint: Look back at your derivation of the expression for ds in part b, bearing in mind that in general, $\sqrt{a^2}$ is not equal to a; it is equal to $|a|$.

38. $3\pi/8$ **39.** a) $(0, 12)$ b) $96\sqrt{3}/5$ units^2 c) $2\int_0^{\sqrt{3}} \sqrt{9t^4 + 46t^2 + 9}\, dt$

40. a) When $t = 1/2$, the point is at $(-11/8,\ 1)$, where the curve's slope is $-16/9$.

b) This occurs when $t = 3$ and when $t = -1/3$, hence at the points $(18, 36)$ and $(26/27\,, 4/9)$. d) $(2{,}4)$

e) 27π units^3. [*Enhanced hint: The volume of a typical horizontal cylindrical slab is* $\pi x^2 dy$. *Now convert to t and integrate, taking care with the boundaries of integration*.]

41. $\sqrt{3}/3$ **42.** $5a^3\pi^2$

Index

About the Author

Seth Braver was born in Atlanta, Georgia. His degrees are from San Francisco State University, University of California (Santa Cruz), and the University of Montana. Before crash landing into South Puget Sound Community College in Olympia, Washington, he taught for brief periods at Portland Community College, the University of Montana, and St. John's College (Santa Fe). His first book, *Lobachevski Illuminated*, won the Mathematical Association of America's Beckenbach Prize in 2015, and has been read by people he has never met. The same cannot be said of his second book, *Ill Enough Alone*, a slim collection of playfully gloomy poetry. When not writing about himself in the third person, Seth mostly minds his own business.

Made in the USA
Las Vegas, NV
01 October 2021

31470216R00105